能源电力英语系列教材
总主编 ● 潘卫民

能源电力工程管理英语翻译教程

English Translation Course for
Energy and Power Engineering Management

主　编　赵德全　刘谦辉
副主编　张晓明　容　庆　苏燕萍
编　委　陈　昊　李姝梦　宁广谊
　　　　宁志敏　徐子茗　刘雅洁
　　　　蒋　雯

上海交通大学出版社
SHANGHAI JIAO TONG UNIVERSITY PRESS

内容提要

本教程分为五章,内容涵盖翻译概论、科技英语的文体特征与翻译、能源电力工程管理的文本特征与翻译、能源电力工程管理英语翻译技巧和能源电力工程管理英语翻译实践。书后还配有能源、电力、工程管理英汉术语表。本教程适合能源电力高校师生和从事能源电力工程管理实践工作的人员阅读使用。

图书在版编目(CIP)数据

能源电力工程管理英语翻译教程/赵德全,刘谦辉主编.—上海:上海交通大学出版社,2024.9—ISBN 978-7-313-31657-8

Ⅰ.TK01;TM

中国国家版本馆 CIP 数据核字第 2024F6X081 号

能源电力工程管理英语翻译教程

NENGYUAN DIANLI GONGCHENG GUANLI YINGYU FANYI JIAOCHENG

主 编:	赵德全 刘谦辉			
出版发行:	上海交通大学出版社		地 址:	上海市番禺路951号
邮政编码:	200030		电 话:	021-64071208
印 制:	上海万卷印刷股份有限公司		经 销:	全国新华书店
开 本:	787mm×1092mm 1/16		印 张:	15.75
字 数:	382千字			
版 次:	2024年9月第1版		印 次:	2024年9月第1次印刷
书 号:	ISBN 978-7-313-31657-8			
定 价:	69.00元			

版权所有 侵权必究

告读者:如发现本书有印装质量问题请与印刷厂质量科联系

联系电话:021-56928178

前　言

近年来,随着国家"一带一路"倡议的实施,中国能源电力项目正大步走向国际化。本教程的出版目的是为电力相关专业大学生提供项目管理基本知识和相关翻译知识技能,满足其紧跟能源电力项目管理的国际化步伐,促进其成为能源电力项目管理与专业翻译的复合型人才。

本教程的编写原则是:一是满足我国能源电力高校的英语专业本科生和研究生群体使用需要。根据教育部相关文件,能源电力高校的英语专业学生在掌握英语专业的基本知识和语言技能的前提下,还应掌握能源电力相关领域的基础知识。能源电力项目管理是英语专业学生学习最多,也是较容易掌握的课程知识之一。因此,我们将该部分知识与能源电力英语翻译结合在一起,编写成本教程,以满足高校培养复合型英语人才的课程需要。二是满足在职电力项目管理人员提高外语尤其是能源电力英语翻译水平的需要。目前我国有大量能源电力工程走向世界,集管理和翻译技能于一身的复合型人才需求量巨大。

本教程分为五章,内容涵盖翻译概论、科技英语的文体特征与翻译、能源电力工程管理文本特征与翻译、能源电力工程管理文本翻译技巧和能源电力工程管理翻译实践。本教程首先从翻译概论开始,目的是让学习者对翻译这一学科有个宏观了解。其次介绍科技英语的文体特点与翻译,因为能源电力英语无疑隶属于科技英语文体范畴,探讨能源电力英语翻译是离不开科技英语文体规律。再次,第三、四、五章以能源电力英语为重点语料,进行翻译技巧和翻译规律的讲解,并附有一定的练习。值得强调的是,第五章为学习者提供了大量的能源电力工程相关的英语语篇,目的是让学习者在阅读理解的基础上,利用前文介绍的翻译技巧进行有效的翻译实践。最后,我们还在附录里总结了能源、电力、工程管理英语常用术语及翻译,以方便学习者查阅。

本教程编写人员有上海电力大学外国语学院赵德全教授、中国电子系统工程第四建设有限公司总裁刘谦辉博士,上海电力大学外国语学院张晓明副教授、苏燕萍讲师、宁志敏副教授、李姝梦老师和研究生徐子茗、刘雅洁、蒋雯，其他编写人员还有容庆副教授、陈昊老师和承德医学院宁广谊讲师。本书的编写得到了上海电力大学经济与管理学院杨太华教授的大力支持，上海电力大学外国语学院翻译团队全体成员也提供了无私支援，在此一并真诚感谢。

目 录

第一章 翻译概论 ··· 001
  第一节 中国翻译简史及翻译思想 ·· 001
  第二节 西方翻译简史及翻译理论 ·· 018

第二章 科技英语的文体特征与翻译 ·· 028
  第一节 科技英语文体 ·· 028
  第二节 科技英语翻译原则 ·· 031
  第三节 科技英语翻译技巧 ·· 034

第三章 能源电力工程管理的文本特征与翻译 ··· 054
  第一节 能源电力工程管理文本特征 ··· 054
  第二节 能源电力工程管理翻译的标准 ·· 060
  第三节 对译者的要求 ·· 062

第四章 能源电力工程管理英语翻译技巧 ··· 072
  第一节 词汇翻译视角 ·· 072
  第二节 句子翻译视角 ·· 082
  第三节 语篇翻译视角 ·· 096

第五章 能源电力工程管理英语翻译实践 ··· 101
  第一节 能源电力工程管理基本知识 ··· 101
  第二节 新时代的中国能源发展 ·· 163

附录 能源、电力、工程管理英汉术语表 ·· 216

参考文献 ··· 244

第一章 翻译概论

了解中西翻译简史及知晓中西有影响力的翻译理论是翻译专业学生的必修课。因此，本教程也毫不例外地从翻译概论开始。本章分别简要介绍中国翻译史及翻译思想和西方翻译史及翻译理论。

第一节 中国翻译简史及翻译思想

中国翻译史有史籍记载的已长达三千余年，早在史前三皇五帝时代就存在翻译活动（中国翻译史话）。马祖毅认为，五四运动（1919年）前有三次翻译高潮：东汉至唐宋的佛经翻译；明末清初的科技翻译；鸦片战争（1840年）以后的西学翻译。本教程分三个时间段介绍中国翻译史，第一时段包括发生在中国古代史和近代史上的这三次翻译高潮；第二时段指五四运动至新中国成立期间的国内翻译；第三时段指新中国成立到改革开放之初的国内翻译。每一时段都包含该时段的翻译活动特点、代表人物及其翻译思想或翻译特点。

一、古代和近代出现的三次翻译高潮

中国古代和近代史上出现的三次翻译高潮在本质上反映了中国历史的发展进程，并对历史的发展和文化的演变起到了重大作用。

1. 第一次高潮：东汉至唐宋的佛经翻译

第一次高潮是东汉至唐宋时期，佛经翻译盛行。我国的佛经翻译从东汉桓帝安世高译经开始，魏晋南北朝时有了进一步发展，到唐代臻于极盛，北宋时已经势微，元代以后则是尾声了（马祖毅，2004：19）。关于佛经翻译，做出较大贡献的人物有安世高、支娄迦谶、支谦、释道安、鸠摩罗什和玄奘。佛经翻译大致分为四个时期：东汉至西晋（草创时期）；东晋至隋（发展时期）；唐代（全盛时期）；北宋（结束时期）。

第一个时期是东汉至西晋，是佛经翻译的草创时期。这一阶段始于东汉桓帝建和二年（公元148年），主要是外籍僧人和华籍胡裔僧人，翻译主要靠直译，甚至是"死译""硬译"，采取口授形式，因此可信度不高。最早的翻译家有两人：安清（即安世高）和娄迦谶（又名：支娄迦谶，简名：支谶）。支谦所作《法句经序》，一般认为是现存最早带有佛经翻译理论性质的文章。然而在当时，他们的译作都没有产生大的影响。究其原因，是汉代的佛教始终被视为当时社会上所盛行的神仙方术迷信的一种。汉代人对佛教的理解，可以说就是道术，他们总是

把"浮屠"和"老子"并称。

安世高是东汉桓帝年间的高僧,史学家认为他是中国最早进行佛典汉译的人。他译了《安般守意经》等三十五部佛经,开后世禅学之源,其译本"义理明晰,文字允正,辩而不华,质而不野"(梁皎慧,高僧传)。安世高主要偏于直译。继安世高之后译经的是支娄迦谶,其所译经典,译文流畅,但力求保全原来面目,"辞质多胡音",即多音译。在中国佛经翻译史上,一直存在"质朴"和"文丽"两派。继安世高、支娄迦谶之后的又一译经大师支谦"颇从文丽",开创了不忠实原著的译风,对三国至西晋的佛经翻译产生了很大的影响。翻译中的"会译"(即将几种异译考校对勘,合成一译)体裁,以及用意译取代前期的音译,也均由支谦始。

第二个时期是东晋至隋,是佛经翻译的发展时期。这一时期最著名的翻译家有(释)道安、鸠摩罗什、真谛、彦琮。道安是中国翻译史上总结翻译经验的第一人,他提出"五失本,三不易"之说。五失本(five deviations from the original),指佛经从梵文译成汉语时,在五种情况下允许译文与原文不一致;三不易(three difficulties in translation)指翻译佛经在五种情况下会使译文失去原来面目,有三件事决定了译事是很不容易的。佛经翻译须合乎原文本意,主张"尽从实录,按本而传,不令有损言游字",明确提出了"敬顺圣言,了不加饰"的直译原则,主张严格的直译。鸠摩罗什提倡意译,主张在存真的原则指导下,不妨"依实出华",讲究译文的流畅华美,因此他所译的佛经都富于文学趣味,一直受到中国佛教徒和文学爱好者的广泛传诵。他虽然倾向意译,但在实践上基本仍然是折中而非偏激的。真谛三藏到中国后20余年适逢兵乱,于颠沛流离中仍能译出一百多部重要经论,是鸠摩罗什以后玄奘以前贡献最大的译师。彦琮在《辩正论》中批评了历代译经之得失,提出"宁贵朴而近理,不用巧而背源",也是坚持忠实第一和倾向直译的。其最大贡献是提出了"八备"(eight qualifications),即一个合格的佛经翻译工作者应该具备的八项条件。

另外,这一时期佛经翻译由私人事业转入了译场翻译,释道安在朝廷的支持下首创译场制度,采用"会译"法来研究翻译。释道安晚年时请来天竺人鸠摩罗什,鸠摩罗什继道安之后创立了一整套译场制度,开集体翻译、集体审校的先河。如前文所述,罗什倾向意译,其译经重视文质结合,既忠实于原文的神情,读来又妙趣盎然。他反对前人译经时用"格义"(用中国哲学的传统概念比附和传译佛学概念)的方法,创立了一整套佛教术语。

第三个时期是唐代,是佛经翻译的全盛时期。我国佛经翻译的四大译家——鸠摩罗什、真谛、玄奘、不空——有两名都出现在唐代(四人中最突出的是罗什和玄奘)。这里我们应该注意隋代的翻译家彦琮。隋代历史较短,译经不多。其中彦琮提出"八备",即做好佛经翻译工作的八项条件,在我国译论史上最早较全面地论述了翻译活动的主体——翻译者本身的问题。到了唐代,佛经翻译事业达到顶峰,出现了以玄奘为代表的大批著名译者。

玄奘即通称的"唐三藏"或"三藏法师",他于贞观二年(公元628年)远度印度学佛求经,17年后归国。他带回佛经657部,主持了比过去在组织制度方面更为健全的译场。在19年间译出75部佛经,共1235卷。他不但把佛经由梵文译成汉文,而且把老子著作的一部分译成梵文,成为第一个把汉文著作介绍到国外的中国人。也是我国历史上促进中印友好和文化交流的首屈一指的人物。玄奘针对佛教术语在翻译过程中出现的意义失真情况,提出了著名的"五不翻"原则。所谓不翻,就是用音译,音译即不翻之翻。"五不翻"指用音译的五种情况(秘密故、含义多故、此无故、顺古故、生善故),即在翻译过程中这五种情况不宜译为汉语,只可用音译的办法来处理。另外,玄奘还总结出来了六种翻译技巧:补充法、省略法、变位法、分合法、译名假借法和代词还原法。这些方法不仅适用于佛经翻译,其原理在普通翻译中沿用至今。

第四个时期是北宋,是佛经翻译的衰落与结束时段。唐代末年,无人赴印求经,佛经翻译事业逐渐衰微。公元907年唐朝灭亡,中国封建社会进入五代十国时期。到了宋代,虽也有人西去求经,印度也有名僧东来传法,宋太宗也曾兴建译经院,从事佛经翻译,但其规模与水平已远不如唐朝的玄奘时期。元、明、清三代从事佛经翻译的人数渐少,几百年间只译了几十部经卷。宋代虽设译经院,但像唐朝时期的大型译场早已不见,对佛教的贡献无法与唐代相比。佛教在印度的衰落导致我国的佛经翻译活动从11世纪开始迅速衰落,译场时代也随之结束。

佛经翻译高潮过去以后,除少数民族地区以外,没有较大规模的文字翻译活动。但各民族的翻译活动在创造、繁荣和发展中华民族文化的过程中也起到过一定的作用。如《元秘史》《古兰经》等的翻译。

佛经翻译对中国语言与文化产生了巨大影响,促进了汉语构词方式的发展,丰富了汉语的音韵、语法、文体和句法,以无法预想的各种方式丰富着中国的文学题材。

2. 第二次高潮:明末清初的科技翻译

在这段时间里,欧洲的一批耶稣会士相继来华进行翻译活动,主要以传教为宗旨,同时也介绍了西方学术。明末清初,西方传教士带来的西学,主要内容是基督教义、古希腊哲学以及17世纪自然科学的若干新发现。这次翻译高潮从延续时间及译著数量上都比不上先前的佛经翻译,但其最重要的成就就是翻译了一些天文、数学、机械等自然科学著作。这一阶段的代表人物主要为中国科学家徐光启和意大利人利玛窦。他们二人合作翻译了著名的《几何原本》前六卷。徐光启是我国明末的杰出科学家、翻译家、进步思想家和爱国政治家。他最早将翻译的范围从宗教、文学扩大到自然科学。后来,李善兰、徐寿、华蘅芳及外国人傅兰雅、伟烈亚力等也有大量译著出现。这一阶段翻译多为外国人口译、中国人笔述,国人选择译品的余地不大,而口译和笔述者对翻译理论与技巧又知之不多,所以译作大都有"文义难精"之弊。但是他们翻译的大量西方科技书籍在普及西方科技知识方面的作用是不能抹杀的。其中,西士傅兰雅在《江南制造局翻译西书事略》中总结出三条科技名词翻译的原则,颇有影响。语言学家马建忠提出了"善译"的翻译标准,即译者必须精通原文和译文,比较异同,掌握两种语言的规律,译书之前,必须透彻了解原文,达到"心悟神解"的地步,然后下笔,忠实地表达原义,"无毫发出入于其间",而且译文又能够摹写原文的神情,仿效原文的语气。这些要求是很高的,都有一定的道理,但由于他本人专门研究语法而没有做翻译工作,因此他对"善译"的见解,被后人忽略了。

3. 第三次高潮:鸦片战争至五四运动的西学翻译

这段时间国内的翻译以西方思想和文学翻译为主,最引人瞩目的就是严复和林纾。严复在《天演论·译例言》中首次提出"信、达、雅"的翻译标准,"信"是"意义不倍(背)本文","达"是不拘泥于原文形式,但严复对"雅"字的解释今天看来是不足取的。他所谓的"雅"是指脱离原文而片面追求译文本身的古雅。他认为只有译文本身采用"汉以前字法句法"——实际上即所谓上等的文言文,才算登大雅之堂。由于时代不同,严复对"信、达、雅"翻译标准的解释有一定的局限性,但许多年来,这三个字始终没有被我国翻译界所废弃。原因是作为翻译标准,这三个字的提法简明扼要,层次分明,主次突出;三者之中,信和达更为重要,而信与达二者之中,信尤为重要。因此有些翻译工作者仍然沿用这三个字作为当今的翻译标准,但旧瓶装新酒,已赋予其新的内容和要求。例如,他们认为"雅"已不再是严复所指的"尔雅"和"用汉以前字法句法",而是指"保存原作的风格"问题。总之,严复对我国翻译事业是有很大贡献的。

林纾他自己不懂外文,主张意译,而且有时他的意译有些过,因此,作品少有人问津。但其所译小说语言优美,语言有时读起来比原作还要好,对文学界和社会风尚产生了很大影响。许多文学家,如鲁迅、郭沫若等也都受到林译小说的影响。

二、五四运动至新中国成立期间国内翻译

五四运动以后,中国翻译理论史进入了现代翻译史。中国译坛又陆续涌现出一大批翻译实践和翻译理论大家,其中有很多都是著名的文学家,如鲁迅、胡适、林语堂、茅盾、郭沫若、瞿秋白、朱生豪、朱光潜等。

1. 翻译特点

五四运动以后,国内的翻译出现了如下特点:①去除了过去"豪杰译",任意增删等弊病,使译文更切合于原文,大大提高了翻译质量;②自推行新文化运动以后,译文大多采用白话文;③从事于翻译者多属留学回国的人员,他们既通外语,又学有专长。特别是从事外国文学翻译的人员,往往又是致力于创作的作家和诗人。而外国文学的翻译又促进中国新文学的创立和发展;④科学翻译着重于外国科学名著的介绍;⑤哲学社会科学的翻译,除译介外国古典文献及资产阶级学者的著作外,又开始系统地译介马克思主义的经典著作。

五四运动是我国近代翻译史上的分水岭。五四运动以前最显著的表现是,以严复、林纾等为代表的翻译者翻译了一系列西方资产阶级学术名著和文学作品。五四运动以后,我国翻译事业开创了一个新的历史时期,开始介绍马列主义经典著作和无产阶级文学作品。《共产党宣言》的译文就发表在五四运动时期。这一时期的翻译白话文代替了文言文。东西方各国优秀文学作品,特别是俄国和苏联的作品,开始由我国近代翻译史上卓越的先驱者鲁迅、瞿秋白等前辈介绍进来。

鲁迅是翻译工作上理论与实践相结合的典范。他和瞿秋白两人关于翻译问题的通信,为我们提供了一些应遵循的基本翻译原则。他译过许多俄国和苏联的优秀文学作品,如《毁灭》《死魂灵》等。他与瞿秋白通过翻译实践,总结了许多宝贵的经验。鲁迅对翻译标准的主要观点是:"凡是翻译,必须兼顾着两面,一当然力求其易解,二则保存着原作的丰姿。"鲁迅竭力反对当时那种"牛头不对马嘴""削鼻剜眼"的胡译、乱译,他针对有人所谓"与其信而不顺,不如顺而不信"而提出了"守信而不顺"这一原则。当然,鲁迅这条原则有"矫枉必须过正"的意味,但与借此作挡箭牌的"硬译""死译"实无共同之处。鲁迅主张直译,是为了保持原作的风貌,这与借直译作挡箭牌的"死译"也丝毫无共同之处。鲁迅和瞿秋白对待翻译工作的态度都十分严肃,至今一直是我们学习的榜样。

这个时期翻译理论开始受到重视并有了长足的发展。翻译论述与批评在此时期出现了百家争鸣的现象。

2. 翻译界代表人物

如上文所述,五四运动前后,我国翻译领域出现了较为繁荣景象,涌现出大批优秀译者,其最具代表性的人物有林纾、严复、鲁迅、胡适、林语堂、梁实秋、瞿秋白、朱生豪、矛盾、郭沫

若、朱光潜等。

1) 林纾（1852～1924）

福建闽县（今福州市）人，是我国近代著名文学家和翻译家。1897年，他凭借《巴黎茶花女遗事》译本开始了他的著译生涯。因为他不懂外语，所以他与朋友王寿昌、魏易、王庆骥、王庆通等人合作，翻译外国小说。他翻译的小说有许多出自外国名家之手，如英国作家狄更斯著的《大卫·科波菲尔德》、英国司哥特著的《撒克逊劫后英雄略》、笛符著的《鲁滨逊漂流记》等。

林纾自幼酷爱中国文学，虽家境贫寒，身处乱世，却不忘苦读诗书；虽为举人（1882年），却因"七上春官，屡试屡败"，最终放弃仕途，转而专心致志走上文学创作之路。林纾作为中国新文化运动的先驱，在其晚年20多年的时间里翻译了180余篇西洋小说，确立了他译界泰斗的地位，也被公认为中国近代文坛的开山祖师，并留下了"译才并世数严林"的佳话，被称为中国新文化先驱及译界之王。

林纾除翻译小说外，也创作了许多小说、散文和诗歌。晚清时期清政府腐败及帝国主义的入侵所导致的民族危机使得立志改革的先进者看到了小说的影响力。林纾的翻译作品大多也是小说翻译，顺应了当时的翻译风潮。从目的论的角度分析，林纾的翻译目的包括两个：政治和文化。政治目的即林纾的爱国精神，他希望通过对西洋小说的翻译能够将新思想传播给中国国民，从思想上挽救落后的中国。因此，他提倡要大量介绍小说，发展翻译事业来"开民智"。文化目的即林纾对中国传统文化的深爱，特别是对文言文的钟爱。林译作品大多是从外文翻译为文言文，而不是当时新文化运动倡导的白话文，这同时体现了林纾后期的顽固保守思想。林纾提出注重翻译的目的与功能，强调译品要能达到预期目标和效果，要求译作能够完成肩负的任务。林纾过分注重其译作中救国思想的宣扬，对原作进行删减和增加，导致一定程度上对原作的不忠实。

林纾翻译中的"讹"：一种是不通西文和粗心大意的"讹"，另一种是明知故犯，创新性的"讹"。第一种"讹"对译文产生荒唐或有害的效果，第二种"讹"则给译文带来了活力。在林纾众多的译文作品中，有一些确实是其滥砍滥伐后的作品，有一些也出现不相关的增加的内容，但其创新性地为原文修枝剪叶，省略了大量与情节发展不相关的，冗长的人物描写分析，再进行润色补充，就大大增强了译文的活力。

忠实论：基本上遵循了严复提出的"信、达、雅"。例如：林纾在翻译中除了因人口译而有差错和删改，一般都能将作者原名列出，书中人名地名绝不改动一音，尤显难能可贵。林译的人名、书名，如大卫·考伯菲尔、耐儿、罗密欧、《鲁滨逊漂流记》《伊索寓言》至今仍在使用。

文风论：林纾精通中西文法，运用自己的文学史观，透视中外作家，剖析外国文学作品，"吃透"作家的"用笔"和"用心"。林纾虽"不审西文"，却能把握原作的精髓。

比较论：林纾指出中外文学共同的表现，即小说的题材、情节、人物可以不同，但其间所表现的情感是相同的。在林纾看来，不管东方西方、中国外国，小说（文学）所传达的情感形式虽有不同，却都具有"共心"的一面。林纾没有认识到白话文为大势所趋，坚持使用古文体翻译外国小说，是他翻译生涯最大的缺陷。倘若林纾能在翻译创作盛期用白话文翻译小说，将会有更多中文基础薄弱的民众读到外国文学。

林纾不懂外文，选择原本之权全操于口译者之手，因而也产生了一些疵误，如把从名著改编或经删节的儿童读物当作名著原作等。即使这样，林纾仍然译了40余种世界名著，这在中国，到现在还不曾有过第二个，称得上是近代翻译界的一个奇迹。虽然受到自身的社会环境、文化习惯、传统思想道德价值观的影响，林纾删节和改写了原作的部分内容，只进行了少量的保留。但总体上说，译者的翻译形式迎合了译文读者的政治需要和文化倾向，译出了一部中国人喜闻乐见的政治小说。

2) 严复（1854～1921）

原名宗光，字又陵，后改名复，字几道，福建侯官县人，近代极具影响力的资产阶级启蒙思想家、著名的翻译家、教育家，新法家代表人物。他先后毕业于福建船政学堂和英国皇家海军学院，曾担任过京师大学堂译局总办、上海复旦公学校长、安庆高等师范学堂校长、清朝学部名辞馆总编辑。在李鸿章创办的北洋水师学堂任教期间，培养了中国近代第一批海军人才，并翻译了《天演论》，创办了《国闻报》，系统地介绍了西方民主和科学，宣传维新变法思想，将西方的社会学、政治学、政治经济学、哲学和自然科学介绍到中国，出版有《严复全集》。

严复所提出的"信、达、雅"的翻译标准，对后世的翻译工作产生了深远影响，是清末极具影响的资产阶级启蒙思想家，中国近代史上向西方国家寻找"真理"的中国人之一。他从光绪二十四年（1898年）到宣统三年（1911年）这十几年间翻译了不少西方政治经济学著作，其中最有影响力的是赫胥黎（T. H. Huxley）的《天演论》（*Evolution and Ethics and Other Essays*）。严复每译一书，都有一定的目的和意义，他常借西方著名资产阶级思想家的著作表达自己的思想。他译书往往加上许多按语，发表自己的见解。严复在参照古代佛经翻译经验的基础上，结合自己的翻译实践，在《天演论》（1898年出版）卷首的《译例言》中提出著名的"信、达、雅"翻译标准。

3) 鲁迅（1881～1936）

中国现代伟大的文学家、翻译家和新文学运动的奠基人。原名周树人，字豫才，浙江绍兴人，出身于破落的封建家庭。青年时代受进化论思想影响。1902年去日本留学，原学医，后从事文艺等工作，企图用以改变国民精神。1909年回国，先后在杭州、绍兴任教。辛亥革命后，曾任南京临时政府和北京政府教育部部员、佥事等职，兼在北京大学、女子师范大学等校授课。1918年5月，首次用"鲁迅"为笔名，发表中国现代文学史上第一篇白话小说《狂人日记》，对人吃人的制度进行猛烈的揭露和抨击，奠定了新文学运动的基石。五四运动前后，参加《新青年》杂志的工作，站在反帝反封的新文化运动的最前列，成为五四新文化运动的伟大旗手。

除了出版大量小说集、杂文集和散文集，鲁迅的一生也翻译了不少外国作品。例如：《工人绥惠略夫》《一个青年的梦》《爱罗先珂童话集》《桃色的云》《苦闷的象征》《热风》《出了象牙之塔》《小约翰》《默》《死灵魂》等。

在翻译风格方面，鲁迅倾向于忠实于原作的直译。鲁迅翻译的很多作品，都能体现出他的直译。从他晚年所译果戈理的长篇小说《死魂灵》可以看出，译文除了个别形容词的出入之外，基本是非常忠实的翻译。相比于前期对苏联文学作品和文学理论的"硬译"来，这部译作句式更加灵活，这也显示了鲁迅对果戈理的文字有更为确切的把握。鲁迅也转译过一些作品，但由于语言层面及其他各种问题，鲁迅的译文中就难免出现了不少很长而且难解的句子。鲁迅非常重视翻译对读者的影响作用，将翻译视为庞大的中国思想库建设、中国新文化建设的事业。

总之，鲁迅的翻译观点可以用概括为：①翻译的目的有两个：一是为革命服务，二是供大家参考;②主张"以信为主,以顺为辅";③以直译为主,以意译为辅。④复译很有必要;⑤反对乱译,提倡翻译批评。

4) 胡适(1891~1962)

安徽绩溪人,原名胡洪骍,字适之。他是我国现代著名学者、诗人、历史家、文学家和哲学家。胡适的翻译思想包含以下几个方面:①翻译的现实意义。胡适认为翻译的目的是为人民服务,是为了输入西方先进文化和先进文学,是为了发挥西方文学名著对中国新文学的借鉴作用。他主张打破旧文学的五言七言格式,创造简单易懂的白话文新文学。纵观其在不同时期的翻译目的:早期是出于爱好和家境需要;中期是面对中央集权和外敌入侵的爱国主义情怀;而在新文化运动时期主要是输入文化,为服务文学革命。②白话文翻译。胡适主张使用白话文,目的是使翻译的西洋名著能够对中国文学起到范文作用,忠于原著,且能表达原著的韵味。③翻译态度。胡适友人的信中多次强调翻译的难处,"翻译是件很吃力的工作,比写作要吃力"。胡适说,我们写文章有三种责任,一是对自己负责,求不致自欺欺人;二是对读者负责,求他们能懂;三是对原著作者负责,求不失原意,不能乱写。即使如此,胡适仍然坚持翻译了许多名著,为后世留下了翻译的范本。④翻译标准是"信"和"达"的统一。胡适主张采用直译的方法,"不应该多费精力去做转译"。有研究表明,自新文化运动以来胡适更重视直译,他主张译者要努力使译文能被大众读懂和接受,是为了借助西方文化改良当时的现代文学。⑤翻译选材。在翻译选材上,胡适主张只译名家名作,他痛斥当时的文学翻译使得充斥着低俗、龌龊的二三流外国作品流入中国。他认为翻译的选材是时代的、具有现实意义的,能够鼓舞人生的向上的作品,是在中国内忧外患的社会背景下能激发民众爱国义气来共同抵御侵略的。

胡适一生翻译了 30 多篇诗歌 17 篇短篇小说等,大多收录在《尝试集》和《尝试后集》中,其中有都德、莫泊桑等人的名著,为中国翻译文学的创作提供了翻译依据,是文学翻译的模范。这无论在翻译理论倡导还是在实践中都对我国的翻译事业做出了巨大贡献。他的书信和文章也是其翻译思想的表达,对推动新文化运动及后期的中国翻译事业起到了重要的指导作用。胡适为古老陈旧的中国文学吹进了一缕科学进步之风,加速了中国翻译文学乃至中国社会的现代化进程。

5) 林语堂(1895~1976)

原名和乐,笔名语堂,福建龙溪人,从小受西方文化的影响,对英语和西方语言学造诣很深。在他的著作《论翻译》中他指出应把翻译作为一种艺术看待。他认为翻译艺术依赖三个条件:一是译者对原文文字及内容上透彻的了解;二是译者有相当的国文程度,能写清顺通达的中文;三是译事上的训练。

作为翻译家,林语堂不仅翻译了许多外国作品,同时还用英语写出了不少脍炙人口的文学作品。林语堂的译作包括《国民革命外记》《女子与知识》《易卜生评传及其情书》《卖花女》《新俄学生日记》等外国作品,直接用英语写的长篇小说有:《京华烟云》《风声鹤唳》《唐人街》《朱门》《远景》《红牡丹》《赖柏英》《逃向自由城》《苏东坡传》《武则天传》等,他的译作和英文小说成功地为中西文化交流做出了非凡的贡献。林语堂就翻译问题曾提出过一些独到的见解,其中影响力最大的就是他提出的关于从心理学角度研究翻译的观点。

在《翻译论》中,林语堂先生谈道:"翻译是门艺术,其成功很大程度上依赖于译者的水平和实践经验,除此之外,再无其他成功的捷径。"林语堂先生把翻译上升到了艺术学派的高度,认为要使译文成为"艺术",译者需要做到三点,一是译者要对原文有清晰透彻的理解;二是译者的中文水平要好,能写出流利畅达的文字;三是严格的翻译训练,并且能在翻译实践中理性认识和学习各种不同的翻译思想。在这个概念上,他还提出了翻译的三条标准:忠实标准、通顺标准和美的标准。

6) 梁实秋(1903~1987)

著名文学评论家、散文家、翻译家。曾与徐志摩、闻一多创办新月书店,主编《新月》月刊。后迁至台湾,历任台北师范学院英语系主任、英语教研所主任、文学院院长、编译馆馆长。代表作有《雅舍小品》《雅舍谈吃》《看云集》《偏见集》《秋室杂文》、长篇散文集《槐园梦忆》等。

作为翻译家,他一生笔耕不辍,译作不断,一生坚持将西方一流文学名著介绍到中国,是我国第一位独立完成《莎士比亚全集》的翻译家。

梁实秋的翻译理论著述不多,其翻译思想散见于其散文及20世纪30年代的翻译论战。他的翻译思想可以概括为:①"信"与"顺"辩证统一的翻译标准。梁实秋强调"信",即"存真",充分忠实原文,尊重原作者。同时他也强调"顺"的重要。他认为,译者可以适时跳出原文句法,使译文更加自然流畅。梁实秋指出,译者应兼顾忠实与通顺,在"信"与"顺"之间做好权衡,尽力再现作品原貌。②谨慎的翻译选择与严谨的翻译态度。首先,他坚持译介世界一流名著。其次,他反对转译。梁实秋曾形象地把原著比作美酒。他认为,译作无论多好,与原著相比,总仿佛是"掺了水或透了气的酒一般,味道多少变了,如果转译,气味就更大了"。最后,梁实秋坚持翻译与研究结合的严谨翻译态度。文学翻译不仅是语言的转换,更与文化紧密相连,翻译态度是保证译作质量的前提。③异化和归化结合的莎翁翻译策略。在语言与文化层面,他更多采用异化策略。例如,翻译人名地名等专有名词时,他不赞成误导读者的中国本土式翻译,主张"音译"。若人名体现人物某种特征时,他会适当归化,在翻译的人名中传达该信息,便于读者了解人物的特征。又如,莎剧中有很多双关语、俚语和俗语,几乎无法翻译,他采用异化策略,然后辅以注释,采用异质文化丰富汉语。梁实秋充分考虑读者,采用异化和规化结合的翻译策略,使译文更加流畅,符合本国读者表达习惯,提高译作的可读性和欣赏性。

7) 瞿秋白(1899~1935)

江苏常州人,翻译了托尔斯泰、果戈里、契诃夫等人的作品,是我国最早直接从俄文翻译俄国文学作品的译者之一,是中国无产阶级革命家、理论家,中国革命文学事业的奠基者,中国共产党早期主要领导人之一。他一生虽然短暂,但却给世人留下了许多不朽的精神财富。在中国翻译的历史长河中,他在翻译及翻译理论方面成就突出,并且具有划时代的、影响深远的意义。

瞿秋白所处的政治历史背景对他翻译思想具有很大的影响,他认为文学是能够影响社会的,在当时的历史背景下,文学可以作为阶级斗争的工具,促进革命斗争的进程。而这也正是他文学创作的出发点。同时,他也认为翻译应该是为无产阶级革命服务的,他认真研究我国文学,并深入到社会生活中,将文学作品的分析评论与翻译结合在一起,推动中国文学的发展。瞿秋白的翻译思想可以概括为:①翻译的标准。瞿秋白主张"忠实、准确",他强调译文必须"既信又顺"。并且瞿秋白还提出"读者概念等同"的翻译标准,这个标准与现代的"等值翻译"和尤金·奈达提倡的"功效对等"可以相提并论,而且比他早了几十年提出。他强调并坚持翻译既要"信"又要"顺"的目的就是要使读者得到"等同的概念"(等同于读原文),从而使译文与原文"功效对等"。在他看来,"顺"和"信"虽是一对矛盾,但矛盾的两个方面可以相互转化,并能有机统一。"信"就是"忠实",而"顺"就是"通顺",求信必须"估量每一个字眼",但翻译者要尽最大努力将"不顺"理顺,如果"不顺"那么就会影响到对原文精神的再现,即"信"。总体来说,他主张概念对等,信顺统一,而这也就是在其后来的翻译实践中体现出来的翻译标准。②翻译的方法。根据瞿秋白长期以来的翻译实践和体会,他主张直译,因为这可以使译文"忠实、准确、通顺",能够更好地对读者再现原文信息。瞿秋白的直译是说:"应当用中国人口头上可以讲得出的白话来写",这样既能保证原作的精神,又能使译文"顺"。他认为"直译"要"估量每一个字眼",这种观点与纽马克的直译观点不谋而合。并且瞿秋白的作品都是直译的结果,是用白话文来翻译的,而他的"顺"恰恰就是在说译文必须要以"白话为本位",而绝不能"以文言为本位",也不能以半文不白的语言为本位,否则译文就会走向不顺,这也进一步从侧面表明他的翻译方法为直译,这既遵循了自己严谨的翻译理论与方法,又说明了他革命家、文艺家的身份,这样可以使当时社会环境下的中国人可以读懂译著,而又使译文不失原文之美。

瞿秋白多译马列主义经典及文艺理论著作,这表明他在翻译文本时带有明确的目的性和倾向性。而且当时所处的环境之下,中国需要像他这样的翻译家,正是有他这样的翻译家,才推动了我国革命的发展。在平时的翻译实践过程中,他忠于原则,表达灵活,将"顺"与"信"二者有机结合使得翻译的表现手法时新,并能使读者更加清楚地理解原文。他在翻译时能够为读者着想,注重翻译效果,多用白话文来翻译,能够使读者通过通俗易懂的语言来读懂马列主义,从而达到"改造社会的目标"。

在翻译理论上,瞿秋白还提出过"翻译应当帮助创造出新的中国现在言语"的著名论点,达到表达的细腻和精密。中国当时处于文言文和白话文交替之时,因此,翻译一方面也帮助了白话文的发展,而瞿秋白正是创造了许多新的词语和字眼,既要求译文和原文意思相同又要求译文通顺流畅,并强调译者在翻译过程中应当发挥主观能动性,对原作的表达进行积极大胆的创造,但不能背离原作。他的这种勇于实践、敢于创新的翻译精神对我国传统译学理论影响深远。

8) 朱生豪(1912～1944)

原名朱文森,又名文生,学名森豪,笔名朱朱、朱生等,浙江嘉兴人,是中国著名翻译家、诗人。朱生豪出生于嘉兴南门一个没落的小商人家庭,家境贫寒。1929年进入杭州之江大学,主修中国文学,以英文为副科。1933年大学毕业后,在上海世界书局任英文编辑,参加《英汉四用辞典》的编纂工作,并创作诗歌。其写有诗集多种,但均毁于战火,同时还在报刊上发表散文、小品文。1935年与世界书局正式签订翻译《莎士比亚戏剧全集》的合同。1936年第一部译作《暴风雨》脱稿,到1937年7月先后译出《仲夏夜之梦》《威尼斯商人》《温莎的风流娘儿们》《第十二夜》等喜剧。1937年日军进攻上海,世界书局被占为军营,已交付的全部译稿被焚。后来他重返上海世界书局,仍抓紧时间进行翻译。1941年日军占领上海,朱生豪丢失再次收集的全部资料与译稿。1942年底他补译出《暴风雨》等9部喜剧。直至因劳累过度患肺病去世,他总共完成了31部戏剧的翻译。他是中国翻译莎士比亚作品较早的译者之一,译文质量和风格卓具特色,为国内外莎士比亚研究者所公认。

翻译风格:朱生豪在探寻一种最大程度上翻译出莎剧风格的汉语文体。中国的戏剧是唱,而外国戏剧是说。既然是说,那就万万不可脱离口语。因此,他译出了汉语版莎剧的风格,那便是口语化的文体。这是一种很了不起的文体,剧中角色不管身份如何,都能让他们声如其人;人物在喜怒哀乐的情绪支配下说出的十分极端的话,同样能表达得淋漓尽致。

诗歌翻译方面,莎士比亚的戏剧尽管是素体诗剧,但是在多数场合都有一定的韵脚,抑扬顿挫,富于节奏感和音乐感。朱生豪钟情于英国诗歌,同时他具备深厚的中国古典文学修养和古典诗词的创作才能,阅读他所翻译的莎士比亚的戏剧可以感受到平仄、押韵、节奏等和谐悦耳的效果。

在翻译莎士比亚作品时,朱生豪选择了极具表达力的口语化的文体,恰如其分地再现了剧中角色的不同身份,使读者感受到人物的喜怒哀乐。莎士比亚的语言是极为难懂的,这并不是因为语言古旧,而是因为他除了运用书面语以外,还大量运用了那个时代生动活泼的口语。口头艺术对于一般中下层市民而言,是他们最便于领会、掌握,也是最大众化的艺术。它能适应各种情绪。莎士比亚当年主要是为演出而写作,考虑得更多的是舞台上的演员和剧院里的观众,而不是案头的读者。因此,朱生豪选择了口语化文体进行翻译,拉近了原作与译文读者的距离。然而,口译化译文文体并不妨碍他在选字和用词上追求典雅和哲理,讲究概括和气势。朱生豪总是能恰如其分地把握通俗与典雅的平衡点,给读者以通俗而高雅的美感。但是,他对粗俗语表达的处理却较为极端,他对原文中不甚雅驯的语句作了不少"净化"处理。

总之,朱生豪在翻译时,尽量模仿了古典戏剧,翻译难度极大,但译文非常成功。译文中采用了很多文言小句,风格古雅,与原文意图颇为吻合。且译文中舞台剧感很强。朱生豪的译文高贵与俚语兼有,归化与异化同处,具有很强的演唱性和舞台表演性。朱生豪是中国莎剧翻译界的泰斗,对莎剧在中国的传播作出了巨大贡献,其译本是中国莎译史上的一座里程碑。

9)茅盾(1896~1981)

原名沈德鸿,字雁冰,浙江嘉兴桐乡人。他是中国现代著名作家、文学评论家和文化活动家以及社会活动家,五四新文化运动先驱者之一,我国革命文艺奠基人之一。常用的笔名有茅盾、玄珠、方璧、止敬、蒲牢、形天等。他早年丧父,经常流离失所。1921年加入中国共产党,1927年失掉组织联系。曾任中华人民共和国文化部长、全国政协副主席。

茅盾的主要翻译作品有《工人绥惠略夫》《战争中的威尔珂》《医生》《世界的火灾》《哈克贝里·芬历险记》。茅盾是中国现代著名文学翻译家,他的翻译思想继承和发展了苏俄文艺学翻译学派的基本观点,认为文学翻译的要旨在于传"神"。茅盾倡导的"艺术创造性翻译"论是他翻译思想的最高成就,具有重要意义。茅盾是最早提出"句调神韵"说的论者,这对文学译本的语言运用提出更高标准。语汇、语句和语段在构成文学作品的形貌的同时,也形成了作品特有的神韵,"一篇文章如有简短的句调和音调单纯的字,则其神韵大都是古朴;句调长而挺,单字的音调也简短而响亮,则其神韵大都属于雄壮。"(《译文学书方法的讨论》)也就是说语汇、语句在文章中就像绘画中的线条与色彩决定着艺术作品的整体风格。

茅盾早年提出的"神韵"论就初步体现了文艺学的翻译观,即文学翻译应该像文学作品如实反映现实生活的实质那样来再现原作的艺术特色。茅盾提倡直译,但他的直译更侧重于在译文里忠实地传达原作的主题、思想、形象等,而不是追求简单的形式对等,在"形貌"和"神韵"发生冲突时,应舍"形"取"神"。茅盾认为文学翻译是文学创作的一种形式,这样就把翻译上升到艺术创作的高度,对翻译提出了文艺学美学的要求。在艺术创造性翻译思想的前提下,茅盾提倡翻译与创作并重,并向翻译工作者提出了从事翻译的要求。茅盾的翻译观及其方法论缜密科学,自成体系,是其一生翻译实践的经验总结,对中国翻译事业的发展影响深远。

长期处在苏联现实主义文学及译论的影响下,茅盾的文艺学翻译观逐渐发展,并达到高峰。1954年,茅盾在一篇报告中提出了他著名的"艺术创造性翻译"思想,这是对他翻译实践的最高经验总结。这一思想把传达原作的艺术意境作为翻译的根本任务。茅盾把中国古典美学中的意境这一概念引入了翻译,对翻译提出了最高要求。他认为,最能吸引读者的是原作的艺术意境,即通过艺术形象使读者对书中人物的思想和行为发生强烈的感情。他认为,文学翻译绝非简单地临摹原作而是一个再创作的过程,这一过程甚至难于文学创作,它要求译者具备多方面的素质,因此要慎重对待翻译过程。

茅盾的直译很独特,不同于单纯强调语言形式等忠于原作的传统"直译",而是一种对原作更深层次的忠实。从文艺学角度考虑,茅盾的直译观无疑是更合理的。对于意译茅盾主要是结合译诗来谈的,他反对任意删改原作的意译,强调要保留神韵。

翻译研究中的文化转向为翻译研究打开了新视野,提供了新的理论阐释,也为探讨茅盾的文学翻译思想提供了理论依据。茅盾在1916年至1949年的33年间,翻译了外国诗歌、散文、小说、戏剧等文学作品近200万字,其中大多是弱小民族、苏俄以及东欧国家的文学作品。在翻译实践过程中,茅盾对翻译问题的认识极为深刻,开创了国内翻译界对译者主体性研究的先河。同时,茅盾致力于世界进步文学的翻译和介绍,为实现他那"疗救灵魂的贫乏,修补人性的缺陷"的政治目的,具有十足的功利性。

10) 郭沫若(1892~1978)

现当代诗人、剧作家、历史学家、古文字学家。原名郭开贞,笔名郭鼎堂、麦克昂等,四川乐山人。郭沫若天性聪颖,从小就表现出其极强的语言学习能力。通过长期的学习实践,他在精通德语、日语、英语的基础上,逐步对这些国家的优秀文化作品产生了浓厚的翻译兴趣,由此走上了一条为各国优秀文化搭建"桥梁"的道路。

学界通常将郭沫若翻译介绍外国文学的活动分为三个阶段:萌芽期——五四运动时期,从翻译介绍德国作家歌德的作品开始;发展期——从1924年流亡日本到40年代末,从翻译介绍日本马列主义经济学学者河上肇的《社会组织与社会革命》开始;成熟期——建国后。在三个阶段中,以第二阶段翻译的作品最多,涉及的面也最广,尤其是完成了《浮士德》和《战争与和平》两部文学巨著的翻译。

在具体的翻译方法上,郭沫若主张"意译"和"风韵译"。所谓"风韵译",也可以叫作"气韵译",即指翻译时不仅不能背离原文的意义,而且需要对于原文的风格保持原样。

郭沫若注重译家责任,重视译者主体性,强调译者主观感情投入,呼唤译界良好风气。郭沫若认为在翻译过程中,译者要有正确的动机和高度的责任感,要对作品进行慎重选择,对所译介的外国作家作品有深刻的了解和研究,尤其是要对作家所处的社会环境、时代潮流等进行彻底考察,译者还要有高度的思想水平和广博的文学修养,才能具有驾御作品内容和语言的能力。

郭沫若还特别强调译者自身所具有的相同或相似的生活情趣对翻译文学的重要性。郭沫若认为,译家对译品有选择的权力,对读者有指导教益的责任,翻译的目的是由译家个人确定的。他的翻译动机观是与译者主体性密切相关的。

郭沫若是我国一代文艺宗师。他的创作和翻译在我国"五四"新文学、翻译文学及现代文学中都占有重要地位。总体来看,郭沫若的翻译具有鲜明的特色。毋庸置疑,无论在理论上还是实践上,郭沫若为我国现代文学、现代翻译文学都做出了巨大贡献。

11) 朱光潜(1897~1986)

字孟实,安徽省桐城人,现当代著名美学家、文艺理论家、教育家、翻译家。1922 年毕业于香港大学文学院。1925 年留学英国爱丁堡大学,致力于文学、心理学与哲学的学习与研究,后在法国斯特拉斯堡大学获哲学博士学位。1933 年回国后,历任北京大学、四川大学、武汉大学教授。1946 年后一直在北京大学任教,讲授美学与西方文学。1986 年 3 月 6 日,朱光潜逝世。朱光潜是北京大学一级教授、中国社会科学院学部委员、中国文学艺术界联合委员会委员、中国外国文学学会常务理事。

朱光潜主要编著有《文艺心理学》《悲剧心理学》《谈美》《诗论》《谈文学》《克罗齐哲学述评》《西方美学史》《美学批判论文集》《谈美书简》《美学拾穗集》等,并翻译了《歌德谈话录》、柏拉图的《文艺对话集》、莱辛的《拉奥孔》、黑格尔的《美学》、克罗齐的《美学》、维柯的《新科学》等。为方便研究马列主义原著,他在花甲之年开始自学俄语,更在八十高龄之际写出《谈美书简》和《美学拾穗集》,翻译近代第一部社会科学著作——维科的《新科学》。

朱光潜进行了长期的翻译实践活动,积累了丰富的翻译经验,更为重要的是,他对这些经验进行了总结,提出了宝贵的翻译思想和观点,对于我国翻译事业的发展和翻译理论研究具有重要的指导意义。朱光潜先生精通美学、诗学、文论和心理学,有着良好的语言学背景和深厚的国学背景。他对翻译问题的探讨更多地是从语言学角度入手,如词汇、句法、语义甚至语用学以及诗学等方面。他的翻译思想具有现代语言学基础,在当时还是很有新意和创见的。在《谈翻译》一文中,他曾谈到严复的"信、达、雅",并对"信"这一标准难以实现的原因从语言学角度进行了分析,"严又陵以为译事三难:信,达,雅。"他列举了文字的四种最重要的意义:直指或字典意义(indicative or dictionary meaning)、上下文意义(contextual meaning)、联想意义(associative meaning)、声音美,又指出文字的两种次要意义:历史沿革意义(historic meaning)、习惯语意义(idiomatic meaning)。翻译终归是不同语言间的转换,朱光潜正是回归到语言本体上探讨翻译问题。

三、新中国成立至改革开放之初的翻译

对翻译而言,新中国至改革开放之初的这段时期属于特殊历史时期。由于受社会政治、经济的影响,翻译活动烙上了时代的印记,丰硕的翻译成果也呈现出自身特有的特点。

1. 翻译活动特点

在这个崭新的时代,任何文化活动都得以前所未有地发展,翻译活动亦是如此。这一时代翻译活动特点可以归纳为五点。

(1) 体现社会主义无产阶级利益的"红色"作品译介。新中国成立后,中国政府把目光投向了人民大众最需要了解的"大众文化",即无产阶级文化。俄国的十月革命为中国送来了代表无产阶级利益的马列主义,苏联社会政治书籍和文学作品成为新中国初期国家政治文化的意识形态基础和主要价值参照。此时的译者更多地把目光瞄准了能代表广大人民群众利益的作品,即"红色作品"。这种出版物有利于教育人民和打击"资产阶级"。在"红色作

品"中,马列著作被放在了首位。1953年1月成立了马列著作专门编译机构——中央编译局,它的任务是有系统有计划地翻译全部马列著作,因为这些译作将作为新中国的政治理论基础。在这一时期,一大批人力都投入到了这项工程浩大的任务之中。中央编译局在师哲等人的领导下,迅速组织人力,大规模编译马恩列斯著作,不少"红色"作品随之诞生,如《马克思恩格斯全集》《列宁全集》《斯大林全集》等。在"红色"作品译介方面,著名的翻译家有郭大力(他翻译了《资本论》《国富论》等经典作品)、陈昌浩(组织和领导编译了《马恩全集》《列宁全集》等作品)、张仲实(著名马列著作翻译家,作品曾获毛主席高度赞扬)等人。时至今日,这些翻译家的作品仍然堪称经典之作并广为流传。

(2) 翻译活动致力于激励社会主义公民的苏联文学作品的译介。在新中国刚成立的这段特殊的历史时期,中国对苏联文化几乎是抱着全盘接受的态度,这为苏联文学作品在中国的译介和传播奠定了良好的基础。无数的中国青年痴迷于苏联的文学作品,许多人把奥斯特洛夫斯基的《钢铁是怎样炼成的》、法捷耶夫的《青年近卫军》、波列伏依的《真正的人》中的主人公当作自己学习的榜样。据统计,1949年到1985年,在译成中文的俄文作品中,作者总数已超5 000人。这一时期,苏联文学在中国找到了一片沃土,苏联文学翻译作品空前繁荣,大批学者、知识分子以及高校教师都投入到了这场声势浩大的苏联文学作品译介中,也涌现出很多优秀翻译家,他们是李俍民(1919～1991)、汝龙(1916～1991)、草婴(1923～2015)、巴金(1904～2005)、满涛(1916～1978)、余振(1909～1996)等。

(3) 除苏联外,西方国家的各类作品,也源源不断进入中国。以文学为例:莎士比亚、弥尔顿、彭斯、斯威夫特、哈代和萧伯纳等英国作家的作品,雨果、司汤达、拉伯雷、左拉、福楼拜、莫里哀、莫泊桑及纪德等法国作家的作品,还有歌德、施托姆、海涅、凯特与格拉斯等德国作家的作品,都被翻译成中文。此外,日本的文化作品译介也呈现一片繁荣的态势。虽然这些作品相比起苏联文化作品而言影响并没有那么大,但也为中国的读者敞开了一扇了解世界的窗口。将欧美、日本作品翻译为中文的翻译家有萧乾、曹未风、翁显良、文洁若、刘振瀛、郑永慧、罗大冈等。

(4) 这一阶段的翻译还向西方翻译介绍中国作品,让外国读者从翻译作品中了解中国。这一时期的中翻英(法日等其他主要语言)在选材上仍然带有一定的政治色彩,比如《毛泽东选集》和《毛泽东诗词》的翻译。专门成立了《毛泽东诗词》英译本定稿小组,由钱锺书和叶君健领衔翻译。另外,中国的唐宋诗词、古典文学作品等都被大量翻译成英文,最具代表性的翻译家就是许渊冲先生。他先后将中国四大诗剧《西厢记》《牡丹亭》《长生殿》《桃花扇》译成英文,还英译法译了毛泽东诗词及大量中国古典诗词。

(5) 政治外交活动翻译事业开始形成。新中国建立后,寻求在国际社会取得认可的中国政府开始重视外交战线的形成。随着外交工作的开展,翻译需求日益增加,因此,为国家领导之间的会晤和其他外交活动进行口译的工作和人才也随之受到国家主要领导人的关怀和重视。1964年外交部正式成立翻译处,周总理十分重视外交干部的培养。在这一领域,从事国家主要领导人的外交翻译工作的几位代表之一是章含之,她亲历了尼克松访华、"上海公报"谈判等一系列重大活动;自20世纪70年代初开始,一直到周恩来、毛泽东辞世以前,唐闻生和外交部的另一位风云人物王海容作为优秀的翻译,几乎参加了这两位伟人与来访各国政要、知名人士的所有会见;还有过家鼎,自1952年从复旦大学毕业后,参加外交谈判、大使级会议,以及陪同周恩来、刘少奇、陈毅、邓小平等国家领导出访,担任外交活动的翻译工作。这些人也正是通过从事高级别外交翻译工作，后来多数成为了外交官。在新中国成立到改革开放之初的这一段特定的历史时期，翻译显示出了空前的繁荣，它虽然带着时代的印迹，但是我们仍能从它的丰富多彩中感受翻译的魅力和力量。同时，它也为改革开放后的翻译之路奠定了良好的基石。

2. 翻译界代表人物

新中国的成立后，翻译事业有了较大的发展，翻译事业如火如荼，涌现出一批又一批的优秀翻译工作者，他们为我国的翻译事业作出了不可磨灭的成绩。比较有影响的翻译工作者有傅雷、钱锺书、王佐良、许渊冲、杨宪益等。有大批学者提出了非常有建树的翻译思想：刘重德提出"信达切"翻译原则；董秋斯在1951年《论翻译的理论建设》中，倡导建立中国翻译学；谭载喜提出建立"翻译学"；罗新璋提出我国翻译"自成体系"说；刘宓庆提出建立"中国特色"翻译学；张南峰提倡"以建立世界翻译学作为中国译界的努力方向"；谢天振强调"中外翻译理论的共通性"；辜正坤提出"翻译标准多元互补论"和"翻译阴阳论"；等等。由于篇幅所限，我们在此只重点介绍钱锺书、傅雷、许渊冲、杨宪益等翻译家及其翻译思想。

1) 钱锺书（1910～1998）

字默存，号槐聚，曾用笔名中书君，江苏无锡人。幼乘家学，饱读经史典籍。1929年考入清华大学外国语文系，1935年考取公费，赴牛津大学英国语文系攻读两年，后转入巴黎大学进修法国文学。1938年归国，先后任西南联大外文系教授、湖南蓝田师范学院英语系主任、上海暨南大学外语系教授和中央图书馆英文总纂。1949年任清华大学外文系教授。1953年起任文学研究所研究员，1982年后任中国社会科学院副院长，院特聘顾问。钱锺书先生学贯中西，在文学等方面造诣甚高，是我国当代著名的人文大师。

钱锺书用"诱""讹""化"三字概括他的翻译思想。翻译在文化领域所起的作用，即诱导人认识外国文学。所谓"诱"也就是我们通常所说的"媒"。钱锺书认为，"媒"和"诱"所指的是翻译在文化交流中的作用。也就是说翻译是个"居间者"和"联络人"，诱使我们去了解外国作品，去阅读外国作品。"讹"主要指的是翻译作品中的"讹错"。译文和原作之间总有"失真"和"走样"的地方，在意义理解或语气上也和原文的风格不符。翻译既是"发现之杖"又是"发现之障"。译文是引领读者进入另一语言文化的助手，然而，也可成为认识另一种语言文化的障碍。钱锺书先生认为正是"讹错"起到了抗腐作用。文学翻译的最高标准是"化"，把作品从一国文字转成另一国文字，既能不因语文习惯的差异而露出生硬牵强的痕迹，又能完全保存原有的风味，那才能说算得上入于"化境"。"化"即是转化（conversion），即上文所述的"将一国文字转成另一国文字"；"化"也是归化（adaptation），将外文用自然而流畅的本国文字表达出来；"化"即是"化境"，也即是"原作的'投胎转世'，躯壳换了一个，而精神姿致依然故我"。这就是钱锺书的"化境"说，与傅雷的"重神似不重形似"和以译莎士比亚名闻遐迩、成绩卓著的朱生豪的"保持原作之神韵"有异曲同工之妙。

语言是认识世界的工具，却又给认识世界制造了最大的障碍，我们常常把语言上的障碍转化为观念上与行为上的障碍。因此，"翻译"是对文化的"沟通"、文化的"对话"便有了特殊的意义。钱锺书用"诱""讹""化"三字高度概括了翻译的性质、功用、易犯的毛病和理想的最高境界，其评论在当代中国译学界独树一帜。同时，从钱锺书对翻译思想的论述中我们也可以深知他的动态、多元、开放的文化观，这种思想为当今全球化语境下进行跨文化传通和文化转型提供了颇具借鉴价值的思路。译文是语言文化灵魂的传递，好的译文是"发现之杖"，而差的译文却是"发现之障"。

2）傅雷（1908~1966）

字怒安，号怒庵，上海人。我国著名的翻译家、作家、教育家、美术评论家、文学艺术评论家。1928年傅雷留学法国巴黎大学，学习艺术理论。开始受罗曼·罗兰影响，热爱音乐。1931年傅雷回国任教于上海美术专科学校，致力于法国文学的翻译与介绍工作。傅雷学养精深，对美术及音乐理论与欣赏等方面有很高的造诣。其翻译思想包括"重神似不重形似""行文流畅，用字丰富，色彩变化""以艺术修养为根本""化为我有""译者修养""翻译态度""神似说"等，其中"神似说"是他翻译思想的核心。

关于"神似说"提出的背景。首先，应是建立在他具和谐美特质的文艺思想上的。傅雷具和谐之美文艺思想的特质，主要表现在对美与善的和谐统一、内在与外在的平衡发展、内容与形式完美结合的追求上。其次，傅雷翻译论提出的美术背景。还在傅雷正式提出"神似"说翻译论前，傅雷就有关于绘画方面的形与神关系的论述，在绘画方面傅雷十分强调神似，他将黄宾虹视为实践绘画"神似"说的杰出代表。关于形与神的关系，傅雷这样写道："山水乃图自然之性，非剽窃其形。"傅雷将他在绘画领域里的"重神似不重形似"的艺术主张贯通在翻译领域，如他说"以效果而论，翻译应当像临画一样，所求的不在形似而在神似"。

"神似"包括两个层面的意思：其一是指追求传达出原作字里行间的涵义和意趣；其二是指追求透出贯穿原作的神韵和风格。而他所谓的"形似"则是指译者在翻译时，最大限度地保留原文的形式，如保留原文的体裁、句法构造、文法和修辞格律等。有的译者说傅雷"神似说"翻译论太过"重神轻形"，但傅雷的"重神似而不重形似"，并不是说只考虑"神似"而把"形似"抛在一边。他曾经说道："我并不是说原文的句法绝对可以不管，译者应该在最大限度内保持原文的句法。"因此，傅雷先生所说的"神似"并不是抛开"形似"，而是把两者都考虑进去。事实上，傅雷也不可能为了传达原作之神韵而完全弃形，因为"形具而神生""形灭而神息"，更何况"形神不和谐"的翻译观背离了他的和谐主义的审美理想。傅雷的"神似"说把翻译纳入文艺美学的范畴，是我国传统译论中不可忽视的重要一笔，将翻译理论推向了新的发展阶段。

傅雷强调形似与神似和谐统一。傅雷将翻译过程分为"译前""译中"和"译后"三个阶段。在"译前"阶段，傅雷强调要审慎选择原作。傅雷选择原作的标准是：原作要与译者性格相近，气质相投，译者喜爱。他以自己选择罗曼·罗兰的原作翻译为例说："自问最能传神的是罗曼·罗兰，第一是同代人，第二是个人气质相近。"其次，要充分吃透原著，将原著的细节了然于心，将原著的精神"化为我有"。为做到充分吃透原著，傅雷提倡"事先熟读原著，不厌其烦尤为要著。任何作品，不精读四五遍决不动笔，是为译者基本法门。第一要将原作（连同思想、感情、气氛、情调等等）化为我有，方能谈到迻译。"；"译中"阶段，强调译者在翻译中传递原作之神韵，实现神似与形似的和谐统一。傅雷常假定译文是原作者的中文写作，以克服在翻译中不利于传神的因素。傅雷说："假定理想的译文仿佛是原作者的中文写作。那么原文的意义与精神，译者的流畅与完整，都可以兼筹并顾，不至于再以辞害意，或以意害辞的弊病了。"；"译后"阶段，指傅雷在译稿完成之后，以"文章千古事，得失寸心知"的严肃认真态度要求自己，只要他认为有待改进的地方，哪怕是推倒译稿重译也不足惜。他说："文字总是难一劳永逸，完美无疵，当时自认为满意者，事后仍会发现不妥""鄙人对己译文从未满意"。

关于翻译态度，傅雷先生一生视文学翻译为神圣崇高的事业，翻译态度极为认真严肃，并主张译者应该深刻地理解、体会原作，然后忠实动人地再现原作的思想风貌。关于翻译修养，傅雷认为译者必须加强自身的修养，要学识渊博，趣味广，修养高，学杂且精。译者应该尽量选择同自己的气质和风格相近的作品，以便有更强烈的共鸣和更好地再现作者的风格。

3）许渊冲（1921～2021）

江西南昌人。早年毕业于西南联大外文系，1944 年考入清华大学研究院外国文学研究所，1983 年起任北京大学教授。从事文学翻译长达 80 余年，译作涵盖中、英、法等语种，翻译集中在中国古诗英译，形成韵体译诗的方法与理论，被誉为"诗译英法唯一人"。在国内外出版中、英、法文著译 60 本，是中国当代翻译家中一位杰出的代表，他翻译的中国古典诗歌作品，为播中国文化、提高中国优秀文化在世界文学界的影响力做出了巨大的贡献。许渊冲先后将中国四大诗剧《西厢记》《牡丹亭》《长生殿》《桃花扇》译成英文，除此之外，他的英译作品还有《诗经》《楚辞》《论语》《老子》《唐诗三百首》《宋词三百首》《元曲三百首》《李白诗选》《苏东坡诗词选》《毛泽东诗词选》等名著；法译作品有《中国古诗词三百首》《诗经选》《唐诗选》《宋词选》《毛泽东诗词四十二首》等。

在大量的诗歌翻译实践活动中，许渊冲继承和发展了中国传统翻译思想，又提出了诗歌翻译的三原则——"三美论""三化论""三之论"，具有鲜明的时代翻译理论特色，为中国优秀诗歌走向世界做出了不可磨灭的贡献，对中国当代的诗歌翻译也有着重要的现实指导作用。

"美化之艺术，创优似竞赛"这十个字是许渊冲教授的文学理论翻译观。"美"是指音美、意美、形美。许渊冲的"三美"说应该是借鉴于鲁迅在《自文字至文章》中所说的"音美以感心，一也；音美以感耳，二也；形美以感目，三也"；"化"是指等化、浅化、深化。他的"三化"论是把钱锺书先生提出的"文学翻译的最高境界是'化'"中的"化"扩展延伸而得来的；而"之"则指的是知之、好之、乐之。"三之"论来源于孔子在《论语》中说的"知之者不如好之者，好之者不如乐之者"；"艺术"二字来源于朱光潜先生提出的"'从心所欲，不逾矩'是一切艺术的成熟境界"。然后，许渊冲又从郭沫若提出的"好的翻译等于创作"中取了一个"创"字，从傅雷提出的"重神似不重形似"中取一个"似"字，从自己提出的"发挥译语优势"中取一个"优"字，再加上"竞赛"二字，最终把文学翻译总结为"创优似竞赛"五个字。这就是说，创造美是文学翻译的本体论，发挥优势是方法论，神似是目的论，竞赛是认识论。"创优似竞赛"和"美化之艺术"并不矛盾，因为本体论是创造"三美"，方法论的"三化"都要发挥译语优势，目的论的"神似"才能使人知之、好之、乐之，认识论的"竞赛"也是一种艺术。这样总结前人的经验，再联系自己的实践，就把中国学派译论概括为："美化之艺术，创优似竞赛。"这就是说，"三美"（意美、音美、形美）是本体论，"三化"（等化浅化深化）是方法论，"三之"（知之好之之）是目的论，"艺术"是认识论。

简单说来，"三美"是文学翻译的本体论，"三化"是方法论，"三之"是目的论，"艺术"是认识论，而"优"则是"三美"合而为一的本体论，"创"是"三化"合而为一的方法论，"似"是"三之"合而为一的目的论，"竞赛"则是包含在"艺术"中的认识论。"三美之间的关系是：意美是最重要的，音美是次要的，形美是更次要的。也就是说，要在传达原文的意美的前提下，尽可能传达原文的音美；还要在传达原文意美和音美的前提下，尽可能传达原文的形美，努力做到三美齐备"。

4）杨宪益（1915～2009）

祖籍江苏省淮安市，中国著名翻译家、外国文学研究专家、诗人。他于1934年毕业于天津英国教会学堂新学书院，后入英国牛津大学学习英国文学等。1940年回国后，任重庆中央大学教授、贵阳师范学院英语系主任、光华大学英语教授等。1943年在南京任国立编译馆编纂，其间翻译了中国古代史巨著《资治通鉴》。1949年任编译馆接管组组长，后任南京市政协委员。1953年调外文出版社工作。1963年后与夫人戴乃迭女士共同翻译出版了中国古典文学名著《红楼梦》《儒林外史》《鲁迅选集》及许多外国文学作品，被公认为译作之经典，为新中国的翻译事业赢得了世界性的声誉。

杨宪益翻译非常注重准确性，其次还注重译、介结合。"译介结合"是杨先生的治学态度所致。杨先生对历史研究感兴趣，也写过一些历史考证的文章。他在介绍作品的同时，总是会做一些文学研究的工作，对作家本身以及创作背景做相应的介绍，以让外国读者更好地走进作品。这对翻译工作者而言，的确是非常珍贵的学术习惯。此外，杨宪益还特别注重译介的系统性，各个历史时期的代表性作品都有涉猎。

杨宪益和戴乃迭最著名的译本，就是《红楼梦》了。《红楼梦》英文全译本，除了杨戴本外，还有一种译本是英国的霍克斯在1973年的译本，由美国企鹅出版社出版，该版本译为《石头记》。在对一些容易引起理解歧义的地方，霍氏译本采用更多的方法是意译，所以他的译本非常符合英语读者的阅读习惯，在英美世界影响力很大。杨戴译本更多地采用的是直译，一般人多以此认为他们的译本只适合中国人来读。其实，这正是杨宪益的高明之处。

今天的译者，多把翻译只看作两种语言间的转化，其实远非如此，它更多是两种文化间的互转。是把一种文化所特有的生活风俗、价值观和宗教信仰等，用比较直接的方式翻译出来，引起另一种文化的惊异，并因此影响另一种文化，还是转化为另一种文化方便接受的语言与表达样式，使其成为那种文化的一部分？这是杨宪益和霍克斯在文化策略选择上的不同。杨宪益考虑的是，如何把自己民族的文化完整地呈现给一个英语世界，并因此来影响英语世界的文化样式。而霍克斯的策略则为，如何最小程度地惊动本民族和读者的文化感受，把《红楼梦》转化成自身文化的一部分。

由于近百年来，英语世界成为一种世界性的价值和文化标准，所以在中国翻译中，直译成为一种主导力量。翻译者根本不考虑中国读者的语言和文化习惯，一味突出的是文化差异，这使得中国文化即使在当下的汉语环境中，也成为一种弱势文化。可以说当下流行的翻译策略，在某种程度上显露的正是文化被殖民的迹象。而英美国家的主流翻译观，更倾向于意译，对原文多采用了非常保守的同化手段，使译文符合本土的习惯和政治需求。不同文化间的差异在这些译文中被掩盖，文化的陌生感在其中被淡化处理，这其实也是一种文化霸权意识。

在这个问题上，杨宪益等一些老一代翻译家是有清醒认知的。所以杨宪益在翻译《奥德修纪》和《牧歌》时多采用意译，译文有很强的中国味，另一种文化的陌生感被减至了最弱，而在翻译《红楼梦》时，采用的却多为直译手法。在这点上，杨宪益先生显示了极高的文化智慧。

第二节　西方翻译简史及翻译理论

西方翻译史约两千年。这段历史为我们积累了一份宝贵的文化遗产,通过总结西方的翻译经验,批判地吸收西方从实践中总结出来的理论、方法,能提高我们的翻译水平。一般而言,西方翻译史是在公元前3世纪开始。西方最早译作是公元前3至2世纪之间出现的,72名犹太学者在埃及亚历山大城翻译的《圣经·旧约》,即《七十子希腊文本》,但从严格的意义上说,西方的第一部译作是在约公元前3世纪中叶安德罗尼柯在罗马用拉丁语翻译的希腊荷马史诗《奥德塞》。不论是前者还是后者,都是在公元前3世纪问世,因此可以说西方的翻译活动自古至今已有2000多年的历史了。它是整个西方发展史上的一个极其重要的组成部分。

一、西方翻译活动的四个阶段

从历史角度看,西方翻译活动可以分为四个阶段:古代西方翻译、中世纪西方翻译、文艺复兴时期的西方翻译和现代西方翻译;但从繁荣程度角度看,西方翻译出现过六次高潮,本节以历史为基础划分时段,将几次翻译高潮穿插其中进行介绍。

1. 古代西方翻译

西方翻译在历史上前后曾出现过六次高潮,古代翻译阶段出现了两次。第一次翻译高潮为肇始阶段,出现在公元前4世纪末。这时盛极一时的古希腊奴隶社会开始衰落,古罗马逐渐强大起来。但是,当时的古希腊文化仍优于古罗马文化,因而对古罗马有着巨大的吸引力,翻译介绍舌希腊古典作品的活动可能即始于这一时期或始于更早的时期。然而,在公元前3世纪中叶,有文字记录的翻译确已问世。被誉为古罗马文学三大鼻祖的安德罗尼柯、涅维乌斯和恩尼乌斯,以及后来的普劳图斯、泰伦斯等大文学家都用拉丁语翻译或改编荷马的史诗和埃斯库罗斯、索福克勒斯、欧里庇得斯、米南德等人的希腊戏剧作品。

安德罗尼柯于公元前3世纪翻译的拉丁文版《奥德赛》被视为西方翻译史上最早的译作,其后的一些大文学家们也都开始尝试用拉丁语翻译或改写古希腊戏剧作品。这一阶段的翻译活动将古希腊文学介绍到古罗马,促进了古罗马文学的诞生和发展。

这是欧洲也是整个西方历史上第一次大规模的翻译活动,其历史功绩在于它开创了翻译的局面,把古希腊文学特别是戏剧介绍到了古罗马,促进了古罗马文学的诞生和发展,对于古罗马以至日后西方继承古希腊文学起到了重要的桥梁作用。

关于这一阶段的翻译思想,西赛罗(Cicero)(公元前106~公元前43)认为,译者应该像演说家一样,使用符合古罗马语言习惯的语言来表达外来作品的内容;直译是缺乏技巧的表现;翻译应保留词语最内层的东西,即意思;翻译也是文学创作;各种语言的修辞手段彼此相通,因此翻译可以做到风格对等。他提出"解释员"式翻译和"演说家"式翻译,即直译与意译。

第二次翻译高潮涌现于古罗马帝国的后期至中世纪初期,是宗教性质的。在西方,宗教势力历来强大而顽固,基督教教会一向敌视世俗文学,极力发展为自身服务的宗教文化。作为基督教思想来源和精神武器的《圣经》,自然成了宗教界信仰的经典。《圣经》由希伯莱语和希腊语写成，必须译成拉丁语才能为罗马人所普遍接受。因此在较早时期就有人将《圣经》译成拉丁语，到公元 4 世纪这一译事活动达到了高潮，其结果就是出现了形形色色的译本。以哲罗姆于公元 382 至 405 年翻译的《通俗拉丁文本圣经》为钦定本，标志着《圣经》翻译取得了与世俗文学翻译分庭抗礼的重要地位。尤其在罗马帝国和中世纪初期，教会在文化上取得了公断地位，《圣经》和其他宗教作品的诠释和翻译得到进一步加强。随着欧洲进入封建社会，"蛮族"建立各自的国家，宗教翻译便占有更大的市场，《圣经》被相继译成各"蛮族"的语言，有的译本甚至成为有关民族语言的第一批文学材料。

在这一时期，哲罗姆的翻译思想有较大的影响力。他认为，翻译不能始终字当句对，而必须采取灵活的原则；译者应当区别对待文学翻译与宗教翻译，换言之他提倡"文学用意译，《圣经》用直译"，并提出了"正确的翻译必须依靠正确的理解"观点。

2. 中世纪西方翻译

中世纪中期，即 11 至 12 世纪之间，大批叙利亚学者到雅典，把大批古希腊典籍译成古叙利亚语，并带回巴格达。在巴格达，阿拉伯人又将其译成阿拉伯语，巴格达一时成为阿拉伯人研究古希腊文化的中心。接下来在西班牙托莱多，大批翻译家把作品从阿拉伯语译成拉丁语，许多古希腊典籍便是从这些阿拉伯文译本转译过来的，这是史上少有基督徒和穆斯林的友好接触。托莱多成为欧洲学术中心，西方翻译史第三次高潮出现，翻译及学术活动延续达百余年之久，影响是非常深远的。

换言之，在第三次翻译高潮中，托莱多取代巴格达的地位，成为欧洲"翻译院"，直到 13 世纪古希腊原本才开始传入托莱多，人们开始直接翻译古希腊原著，不再转译。

托莱多翻译院的三大特点：①翻译活动始终得到教会的资助；②翻译的作品主要是古希腊作品的阿拉伯语译本，然后才是阿拉伯语原作和希腊原作；③托莱多是当时西班牙的教育中心和穆斯林学术中心。

托莱多大规模翻译的历史意义：①它标志着基督教和穆斯林教之间罕有的一次友好接触；②该高潮带来了东方人思想，传播了古希腊文化，活跃了西方的学术空气，推动了西方文化发展；③由于许多译者同时也是学者，在托莱多讲授各种知识，托莱多成为当时西班牙以至西欧的教育中心，并在某种意义上成为了西班牙中北部地区第一所大学的前身。

3. 文艺复兴时期的西方翻译

14 至 16 世纪欧洲发生的文艺复兴运动，是一场思想和文学革新的大运动，西方翻译出现了第四次高潮，也是西方翻译史上的一次大发展，特别是文艺复兴运动在西欧各国普遍展开的 16 世纪及尔后一个时期，翻译活动达到了前所未见的高峰。

文艺复兴是对古希腊、古罗马文学艺术和科学的重新发现和振兴。以传播人文主义思想为主要表现形式之一。它不仅标志着文学艺术上的大发展，同时也是翻译史上的重要里程碑。在德国，民族自我意识加强，15 世纪模仿拉丁语之风逐渐消失，翻译重点从偏重原文语言转移到重视译文语言，意译法取代了逐词对译法占据了主导地位。宗教改革家马丁·路德顺从民众的意愿，采用民众的语言，于 1522 至 1534 年翻译刊行第一部"民众的《圣经》"，开创了现代德语发展的新纪元。在法国，复古之风开始盛行，翻译中心从宗教作品转向古典文学作品，翻译活动日趋频繁。文学翻译家雅克·阿米欧（1513～1593?）先后用了 17 年（1542～1559 年）时间，译出了普鲁塔克的《希腊罗马名人比较列传》（简称《名人传》），该翻译与查普曼 1598 至 1616 年译的《伊利亚特》，成为法国 16 世纪译作的典范，在文学领域为翻译争得一席之地。理论方面,多雷领先德英各国理论家,首次比较系统地提出了关于翻译的基本原则。在英国,后来者居上,经济发展提供了雄厚的物质基础。特别是16世纪中叶伊丽莎白登位至17世纪初期,宗教翻译方兴未艾,古代希腊罗马和当代其他国家的文学作品大量译成英语,使这一时期英国的翻译活动成为了欧洲及至世界翻译活动的高峰之一。弗罗里欧1603年所译蒙田的《散文集》,是英语文学译著中一颗灿烂的明星。而1611年《钦定圣经译本》的翻译出版则标志着英国翻译史上又一次大发展。它以其英语风格的地道、通俗和优美赢得了"英语中最伟大的译著"的盛誉,在长时期里成为英国唯一家喻户晓、人手一册的经典作品,对现代英语的发展产生了深远的影响。

总之,翻译活动深入到思想、政治、哲学、文学、宗教等各个领域,涉及古代和当代的主要作品,产生了一大批杰出的翻译家和一系列优秀的翻译作品。文艺复兴时期乃是西方(主要是西欧)翻译发展史上一个非常重要的时期,它标志着民族语言在文学领域和翻译中的地位终于得到巩固,同时也表明翻译对民族语言、文学和思想的形成和发展所起的巨大作用。

文艺复兴后,从17世纪下半叶至20世纪上半叶,出现了西方翻译史上的第五次翻译高潮。在这一阶段,西方各国的翻译继续向前发展。虽然就其规模和影响而言,这一时期的翻译不能与文艺复兴时期相提并论,但仍然涌现出大量的优秀译著,其最大特点是,翻译家们不仅继续翻译古典著作,而且对近代的和当代的作品也发生了很大的兴趣。塞万提斯、莎士比亚、巴尔扎克、歌德等大文豪的作品都被一再译成各国文字,东方文学的译品也陆续问世。这段时间出现的翻译家和翻译理论家有德莱顿(Dryden,1631~1700)、泰特勒(Tytler)施莱尔马赫(Schleiermacher,1768~1834)、洪堡(Humboldt,1767~1835)、阿诺德(Arnold,1822~1888)。

4. 现代西方翻译

西方翻译的第六次翻译高潮主要表现在第二次世界大战结束以来的翻译活动。其特点是:①翻译范围不断扩大;②翻译规模也大大超过了以往;③翻译的作用也为以往所不可企及。第二次世界大战后,西方进入相对稳定的时期,生产得到发展,经济逐渐恢复,科学技术日新月异。这是翻译事业繁荣兴旺的物质基础。由于时代的演变,翻译的特点也发生了很大的变化。新时期的翻译从范围、规模、作用直至形式,都与过去任何时期大不相同,取得了巨大的进展。首先是翻译范围的扩大。传统的翻译主要集中在文学、宗教作品的翻译上,这个时期的翻译则扩大到了其他领域,尤其是科技、商业领域。其次,翻译的规模大大超过了以往。过去,翻译主要是少数文豪巨匠的事业。而今,翻译已成为一项专门的职业,不仅文学家、哲学家、神学家从事翻译,而且还有一支力量雄厚、经过专门训练的专业队伍承担着各式各样的翻译任务。再次,翻译的作用也为以往所不可企及,特别是在联合国和欧洲共同市场形成之后,西方各国之间在文学、艺术、科学、技术、政治、经济等各个领域的交流和交往日益频繁、密切,而所有这些交际活动都是通过翻译进行的,因为翻译在其间起着越来越大的实际作用。最后,翻译事业发展的形式也有了很大变化和进步。这主要体现在三个方面:①兴办高等翻译教育,如法国、瑞士、比利时设有翻译学校或学府,英、美、苏等国在大学高年级开设翻译班(translation workshop),以培养翻译人员;②成立翻译组织以聚集翻译力量,最大的国际性组织有国际翻译工作者联合会(简称"国际译联")以及国际笔译、口译协会和各国的译协;③打破传统方式,发展机器翻译。这第三点实际上是新时期发展的一个重要标志。自1946年英美学者首次讨论用计算机做翻译的可能性以来,翻译机器的研制和运用经过近80年的曲折历程,已日益显示出生命力。它是对几千年来传统的手工翻译的挑战,也是翻译史上一次具有深远意义的革命。目前,西方翻译事业仍处于第六次高潮之中。今后向何处发展,第六次高潮会持续多久,眼下尚难预测。

二、西方翻译理论概述

同其他的领域的活动一样,翻译并不依赖理论而存在,然而,翻译的实践总是引出理论并推动理论的进步,理论又反过来指导实践,促进翻译事业的发展。

1. 从历史角度看西方翻译理论

西方最早的翻译理论家是古罗马帝国时期的西塞罗。他首次把翻译区分为"作为解释员"(ut interpres)和"作为演说家"(ut orator)的翻译。西塞罗是从修辞学家、演说家的角度看待翻译的。所谓"作为解释员"的翻译是指没有创造性的翻译,而所谓"作为演说家"的翻译则是指具有创造性、可与原著媲美的翻译。这样,西塞罗便厘定了翻译的两种基本方法,从而开拓了翻译理论和方法研究的园地。自西塞罗以来,西方翻译理论史便围绕着直译与意译、死译与活译、忠实与不忠实、准确(accuracy)与不准确(inaccuracy)的问题向前发展。

继西塞罗之后,西方翻译史拥有一大批优秀的翻译理论家。他们在不同时期,从不同的角度,提出了各种不同的理论和观点。在古代,除西塞罗的直译和意译的两分法外,还有昆体良的"与原作竞争"和哲罗姆的"文学用意译,《圣经》用直译"之说;有奥古斯丁的《圣经》翻译凭"上帝的感召"和他的有关语言符号理论。在中世纪,有波伊提乌的宁要"内容准确",不要"风格优雅"的直译主张和译者应当放弃主观判断权的客观主义观点;有但丁的"文学不可译"论。在文艺复兴时期,有伊拉斯莫的不屈从神学权威、《圣经》翻译靠译者的语言知识和路德的翻译必须采用民众语言的人文主义观点;有多雷的译者必须理解原文内容、通晓两种语言、避免逐字对译、采用通俗形式、讲究译作风格的"翻译五原则"。17至19世纪,有巴托的"作者是主人"(译者是仆人)、译文必须"不增不减不改"的准确翻译理论;有德莱顿的"直译""意译""拟作"的翻译三分法和翻译是艺术的观点;有泰特勒的优秀译作的标准,即译作应完全复写出原作的思想、译作的风格和手法应和原作属于同一性质、译作应具备原作所具备的通顺等翻译三原则;有施莱尔马赫的口译和笔译、文学翻译与机械性翻译的区分;有洪堡的语言决定世界观和可译性与不可译性的理论;有阿诺德的"翻译荷马必须正确把握住荷马特征"的观点。20世纪,有费道罗夫的翻译理论首先"需要从语言学方面来研究"、翻译理论由翻译史、翻译总论和翻译分论三部分组成的观点;有雅克布逊的"语内翻译""语际翻译""符际翻译"的三类别;有列维的"翻译应为使读者产生错觉""翻译是一种作决定的过程"和加切奇拉泽的"翻译永远是原作艺术现实的反映""文艺翻译是一种艺术创作"的文学翻译理论;有弗斯、卡特弗德的翻译在于"语言环境对等"的语言学翻译理论;有奈达的"等同的读者反应"和"翻译即交际"的理论。可以说,所有这些主要观点都是构成西方翻译理论的重要组成部分。

2. 西方翻译理论的两条主线

西方翻译理论主要由两条不同的线构成。一条是文艺学翻译线,这是一条最古老的线,从泰伦斯等古代戏剧翻译家一直延伸到现代翻译理论家列维(捷克)和加切奇拉泽(苏联)。按照这条线,翻译被认为是一种文学艺术,翻译的重点是进行再创造。理论家们除不断讨论直译和意译、死译与活译的利弊外,对翻译的目的和效果也进行了分析。他们强调尊重译入语文化,讲究译文的风格和文学性,要求译者具有天赋的文学才华。另一条是从古代的奥古斯丁延伸到20世纪的结构语言学派,是语言学翻译理论线。它把翻译理论和语义、语法作用的分析紧密结合起来,从语言的使用技巧上论述翻译,认为翻译旨在产生一种与原文语义对等的译文,并力求说明如何从词汇和语法结构上产生这种语义上的对等。这两条线各有其偏颇之处。文艺学翻译理论强调翻译的目的和结果,从宏观上强调译文的艺术效果,不甚研究翻译的实际过程和语言的使用技巧问题,也忽略非文艺作品的翻译和文艺作品翻译中的非创造功能。语言学翻译理论的缺陷是,不甚注意作品的美学功能,忽略文艺作品的艺术再现,理论分析往往局限于单个的词、句子或语法现象,而忽略话语结构这一更为广泛的内容。直到最近十年来,随着语言学和翻译理论研究的深入,这种缺陷才开始有所修正。

从发展的趋势看,语言学翻译理论线已占据现代翻译理论研究中的主导地位。理论家大都认为,翻译属于语言学的研究范围,是应用语言学和对比语言学的研究范围,是应用语言学和对比语言学的一个分支,与语义学有着密切关系,同时又与文艺学、社会学、人类学、心理学、控制论、信息论等多种科学有关。必须看到不论是文艺学翻译理论还是语言学翻译理论,它们乃是相辅相成的。翻译既不是在所有时候都是创造性、文学性的,也不是在所有时候都只是传递客观信息的。尤其重要的是,翻译理论的研究必须与实践紧密结合,以指导翻译实践、揭示翻译活动的客观的和内在的规律为其唯一的目的。否则,翻译理论就会失去它的生命力。

3. 从翻译学派角度看西方翻译理论

关于西方翻译理论的流派,不同学者有不同的划分角度。谭载喜在介绍西方翻译理论时,将西方翻译理论分为四大学派:布拉格学派、伦敦派、美国结构派和交际理论派。而柯平在《西方翻译理论浅析》一文中介绍了六大学派:语言学派、交际学派、美国翻译研讨班学派、文学-文化学派、结构学派和社会符号学派。

在本教程中,我们综合不同学者的划分视角,重点介绍以下五个学派:文艺学派、语言学派、翻译研究学派、阐释学派和解构学派。

1) 文艺学派

文艺学派起源于古代西方翻译活动。主要代表人物有西塞罗、贺拉斯、哲罗姆(古罗马)、德莱顿、泰特勒(17、18世纪英国)。

西塞罗(Cicero)曾说:"我不是作为解释员,而是作为演说家进行翻译的。"他首次提出"解释员式翻译"与"演说家式翻译",即"直译"与"意译"两种基本译法,确定了后世探讨翻译的方向。西塞罗提出,翻译家必须照顾译语读者的语言习惯,用符合译文读者的语言来打动读者或听众;翻译要传达的是原文的意义和精神,并非原文的语言形式;因为文学作品的翻译就是再创作,翻译文学作品的译者必须具备文学天赋或素质;由于各种语言的修辞手段"彼此有相通之处",翻译中做到风格对等是完全可能的(谭载喜,1991)。自西塞罗的观点发表以来,翻译开始被看作是文艺创作。

贺拉斯(Horatius)提倡"忠实原作的译者不适合逐字死译"。这句话经常被翻译家引用,成为活译、意译者用来批评直译、死译的名言。他主张在创作和翻译中不要墨守成规,必要时可以创造新词或引进外来词,以便丰富民族语言和增强作品的表现力。

哲罗姆(Jerome)完成了第一部"标准"拉丁语《圣经》的翻译,并提出了翻译理论和切实可行的翻译原则,即区别对待文学翻译和宗教翻译,提出"文学用意译,圣经用直译"。

德莱顿(John Dryden)明确提出翻译是一门艺术;译者必须考虑译文的读者和对象。他将翻译分为三类:逐字翻译、意译和拟作。认为逐字翻译是"戴着脚镣在绳索上跳舞";拟作近似于创造,脱离了原作的面貌;主张重意义、轻语言形式的意译。他对翻译的三分法突破了传统二分法(即直译、意译)的局限,可以说是西方翻译史上的一大发展,具有重要的启示意义。

泰特勒(Alexander Tytler)的翻译理论和思想主要见于《论翻译的原则》一书。该书是西方翻译理论的第一部专著。他认为优秀的译作必须使读者领略原作的优点,并得到"同样强烈的感受"。他提出了翻译三原则:①译作应完全复写出原作的思想;②译作的风格和手法和原作属于同一性质;③译作应具备原作所具有的通顺。

2) 语言学派

奥古斯丁发展了亚里士多德的"符号"理论,提出了语言符号的"能指"、"所指"和译者"判断"的三角关系,开创了西方翻译理论的语言学传统。20世纪初,索绪尔提出普通语言学理论,标志着现代语言学的诞生,也为当代翻译研究的各种语言学方法奠定了基础。这个学派内部出现了各种不同流派的代表人物和理论方法,但他们都有一个共同的特征,就是以语言为核心,从语言的结构特征出发研究翻译的对等问题。语言学派主要分为布拉格学派、伦敦学派、美国结构派、交际理论派、德国功能派等。

(1) 布拉格学派代表人物为雅各布森。这个流派的主要翻译论点是:首先,翻译必须考虑语言的各种功能;其次,翻译必须重视语义、语法、语音、语言风格及文学体裁方面的比较。雅各布森(Roman Jokobson)作为学派的创始人之一,所著《论翻译的语言学问题》为该派开山之作。文章从语言学的角度,对语言和翻译的关系、翻译的重要性,以及翻译中存在的问题作出了详尽的分析和论述。提出翻译三分法,即语内翻译、语际翻译、符际翻译,认为对于词义的理解取决于翻译,在语言学习和语言理解过程中,翻译起着决定性作用。布拉格首次提出对等概念,提出准确的翻译取决于信息对称,翻译所涉及的是两种不同语符中的对等信息。他认为所有语言都具有同等表达能力。如果语言中出现词汇不足,可通过借词、造词或释义等方法对语言进行处理。语法范畴是翻译中最复杂的问题,这对于存在时态、性、数等语法形式变化的语言,尤其复杂。

(2) 伦敦学派代表人物是卡特福德和纽马克。伦敦学派从社会学角度研究语言,认为语言的意义是由言语使用的社会环境所决定的,语义理论不仅要规定语法范畴和语法关系,而且要说明文化环境对语义情景的影响。反映到翻译领域,该派观点认为译文的选词是否与原文等同,必须看它是否用于相同的言语环境中。卡特福德(Catford)著有《翻译的语言学理论》,从翻译性质、类别、对等、转换、限度等方面着重阐述了"什么是翻译"这一中心问题。(具体见李文革《西方翻译理论流派研究》2004)纽马克(Peter Newmark)从语言意义入手,将翻译定义为:把一种语言中某一语言单位或片段,即文本或文本的一部分的意义用另一种语言表达出来的行为,把翻译定性为既是科学,又是艺术,也是技巧。他区分语义翻译和交际翻译:语义翻译指在译入语语义和句法结构允许的前提下,尽可能准确地再现原文的上下文意义。交际翻译指译作对译文读者产生的效果尽量等同于原作对原文读者产生的效果。把文本类型归纳为表达功能、信息功能和呼唤功能。他提出关联翻译法:原作或译语文本的语言越重要,就越要紧贴原文翻译。对于翻译批评,他就翻译批评的目的、标准、步骤与方法提出了见解,主要体现在《翻译入门》《翻译教程》中。

（3）美国结构主义语言学派代表人物是布龙菲尔德。他提出一种行为主义的语义分析法，认为意思就是刺激物和语言反应之间所存在的关系。20世纪50年代，布龙菲尔德理论为乔姆斯基的转换生成理论所取代。乔氏理论有三个观点：①人类先天具有语言能力；②语言是由规则支配的；③语言包括表层结构和深层结构。该理论对翻译研究的影响主要在于其关于表层结构和深层结构的论述。语言的不同主要在于各自的表层结构不同，而深层结构则具有共同特点。

在上述语言学理论的影响下，形成以沃哲林（C. F. Voegelin）、博灵格（D. Bolinger）、卡兹（J. J. Katz）、奎恩（W. V. Quine）和奈达（E. U. Nida）为代表的美国翻译理论界的结构学派，而以奈达最为杰出。

（4）交际理论派最重要代表人物是奈达（Nida），他也是整个语言学派最重要的代表人物之一，著述极丰，其理论对西方当代翻译研究作出了很大的贡献。他提出了"翻译的科学"这一概念。在语言学研究的基础上，他把信息论应用于翻译研究，认为翻译即交际，创立了翻译研究的交际学派。奈达提出了"动态对等"的翻译原则，并进而从社会语言学和语言交际功能的观点出发提出"功能对等"的翻译原则，这是他翻译思想的核心概念。所谓"功能对等"，就是说翻译时不求文字表面的死板对应，而要在两种语言间达成功能上的对等。就翻译过程，他提出了"分析""转换""重组"和"检验"的四步模式。

（5）德国功能学派。20世纪60、70年代，德国译学界受结构主义语言学的影响，形成了以纽伯特（A. Neubert）、卡德（O. Kade）为代表的莱比锡派（the Leipzig School）和以威尔斯（W. Wilss）为代表的萨尔派（Saarbrücken School）。前者立足于转换生成语法，在翻译中严格区分不变的认知因素与可变的语用因素；后者是奈达学说的追随者，主张建立翻译科学。

功能派翻译理论在这时兴起，针对翻译语言学派中的薄弱环节，广泛借鉴交际理论、行动理论、信息论、语篇语言学和接受美学的思想，将研究的视线从源语文本转向目标文本。目的论影响深远，功能学派因此有时也被称为目的学派。

赖斯（Reiss）是功能派翻译理论创始人。他提出将文本功能列为翻译批评的一个标准，指出翻译批评的依据应是原文和译文两者功能之间的关系。他把文本类型分为：信息型、表达型、操作型，不同的类型应采用不同翻译策略。他认为译文必须连贯一致，而这种连贯性取决于译者对原文意图的理解。赖斯的理论总体上是建立在对等理论之上，其实质指的是译文与原文的功能对等。

弗米尔（Vermeer），功能学派的另一位重要代表人物，提出了目的论（Skopos theory）。该理论包含三个法则：目的法则、连贯法则、忠实性法则。他以行为理论为基础，提出"翻译行为"概念，认为所有翻译遵循的首要法则是目的法则，翻译行为所要达到的目的决定整个翻译行为的过程，即结果决定方法。

诺德（Nord）是功能学派的另一位重要人物，提出了功能加忠诚理论。他将翻译分为工具性翻译和文献型翻译，前者指翻译作为译入语文化新的交际行为中的独立信息传递工具，后者指翻译作为原文作者和原文接受者在源语文化交际中的文献。

3）翻译研究学派

翻译研究学派主要包含霍姆斯的早期翻译研究学派、埃文-佐哈尔的多元系统学派、图里的描写学派，以及文化学派、女权主义、食人主义和后殖民主义。

(1) 早期翻译研究学派代表人物是霍姆斯（Holmes），翻译研究派的创始人，其著作《翻译研究的名与实》是该派的成立宣言。该学派的出现标志着翻译研究三大分支的形成：描写翻译研究、理论翻译研究、应用翻译研究。他首先认为翻译理论应产生于对翻译过程的科学描述，再将理论应用于翻译实践和翻译教学。翻译研究的重点应该是译文与原作作为两种文学自足体之间的关系以及译文与译入文化之间的关系，而不是传统的对等或忠实。

(2) 多元系统学派：该翻译理论产生于早期翻译研究派，是早期翻译研究派的延伸与发展。多元系统派这一名称由佐哈尔于20世纪70年代首先提出，最终成为低地国家（荷兰、比利时、卢森堡）和以色列学者翻译理论与思想的旗帜。

多元系统学派认为应将翻译的理论概念置于更大的文学、社会和文化的框架之中来考察，必须在翻译研究中引进文化符号学。早期翻译研究学派与多元系统学派的区别在于，前者注重翻译的一对一的等值关系，而后者则认为接受文化的社会和文学标准决定了译者的美学假设，因而影响着翻译的全过程。多元系统学派认为文学与超文学世界可以划分为多层结构系统，文学作品是一个系统，社会环境又是另一个系统，它们相互联系，辩证地相互作用，共同协调某一特定形式因素的功能。佐哈尔（Even-Zohar）创立的术语"多元系统"是指在一定文化中始终存在着主要和次要的文学系统，而高雅文学在其中又居于重要地位。在研究翻译文学作品的社会功能时，他指出，翻译文学并非在所有国家均处于无足轻重或边缘的地位。翻译文学作品不仅引进新的思想，而且还提供新的形式和模仿的样板。如果翻译文学在一个民族中处于次要地位，译者就常常牺牲原作的形式，竭力使译文与接受文化的现行标准保持一致。

(3) 描写学派：图里（Gideon Toury）是描写学派的代表人物。他指出任何翻译都不可能与原文完全契合，忠实的标准因此只能是相对的，并引入了"翻译规范"的概念，强调译语文化规范对翻译策略选择的影响。图里致力于描写译学的构建和方法论的探讨，认为翻译理论的主要任务不是评定译文，而是阐述译文形成的过程，发现目标文化系统的文学趋向对译文的影响。他构建了描写译学，为西方译论奠定了良好的方法论基础。

(4) 文化学派：该翻译学派的两个重要代表人物是勒弗维尔和巴斯奈特。勒弗维尔把翻译看作"重写"，重写就是"操纵"。他提出了翻译研究中"文化转向"问题，即翻译研究不应局限于翻译本身，而应把翻译看作是一种文化发展的策略来研究。勒弗维尔探讨翻译与意识形态、诗学等的关系，提出"翻译实际上是文化融合"。

第二位代表人物巴斯奈特指出，翻译绝不是一个纯语言的行为，它深深根植于语言所处的文化之中，翻译就是文化内部与文化之间的交流。翻译等值就是原语与译语在文化功能上的等值。他的翻译思想可归纳为四点：①翻译应以文化作为翻译单位，而不应停留在语篇之上。②翻译并不只是一个简单的解码—重组过程，更重要的还是一个交流的行为。③翻译不应局限于对原语文本的描述，而在于该文本在译语文化里功能的等值。④不同时期翻译有不同的原则和规范，翻译就是满足文化的需要和一定文化里不同群体的需要。

(5) 女权主义翻译理论：与传统译论对"一致性"的诉求相反，女性主义翻译致力于"差异的凸显"。"差异"有两层含义：一是指不拘泥于与原作的一致；二是强调女性话语与男性话语的差异。这样的思想一方面抛弃了传统的以原作为中心的翻译理念，肯定了影响翻译行为的各种因素；另一方面借翻译的名义，构建女性话语的壁垒。女权主义翻译理论流行于加拿大，代表人物为谢莉·西蒙（Sherry Simon），她著有《翻译的性别：文化等同和传递政治》,考察了性别与翻译之间相互作用的一个特定领域。

(6) 后殖民主义翻译理论:最主要的理论依据生成于20世纪60年代中期西方思想界内部,即以反叛、颠覆二元对立的西方哲学传统和文学批评话语为特征的解构主义。"翻译是帝国的殖民工具"成了后殖民视角下的翻译研究的一个重大命题。该理论试图超越语言的局限,从政治、权力等视角探究翻译问题。它关注的并非翻译本身,但给传统的翻译观念带来极大冲击。后殖民主义理论把翻译和政治联系在一起,探讨弱小民族语言的文本被翻译到强大民族的语言时,译者因意识形态、权力等因素的影响而采取的不同翻译策略。

4) 阐释学派

阐释学派总体翻译观为:译者不是消极地接受文本,而是积极地创造文本的过程;强调翻译和理解之间密不可分的关系,对理解的作用及方式进行不同的阐述。

该学派代表人物之一为施莱尔马赫。他提出了翻译的两个途径,即译者可以"不打扰原作者而将读者移近作者"——以作者为中心;"尽量不打扰读者而将作者移近读者"——以译文读者为中心。施莱尔马赫首次提出翻译应区别口译与笔译,区分了真正的翻译和机械的翻译,前者是指文学作品和自然科学的翻译,后者指的是实用性的翻译。

斯坦纳是阐释学派的另一代表人物。他著有《通天塔:语言与翻译面面观》。其主要观点是:理解即是翻译。该著作中提出了阐释学的翻译步骤:信任、攻占、吸纳、补偿。即在实际翻译过程中,译者不可避免地将个人的生活经验、文化和历史背景渗入了原文,使翻译变成了对原文的再创造。

5) 解构学派

解构主义思潮是20世纪60年代后期在法国兴起的一种质疑理性、颠覆传统的全开放式的批判理论,它以解释哲学作为哲学基础,主张多元性地看问题,旨在打破结构的封闭性,颠覆二元对立的西方哲学传统。解构学派翻译理论强调消除传统的翻译忠实观,突出译者的中心地位。该学派中最著名的学者有德里达和韦努蒂。

(1) 德里达的翻译观:德里达认为,翻译应该重新定义。即翻译不仅仅只定义为掌握某种内容的跨越活动,同时更主要的是为"播散"和"逃遁"(escape)等跨越时间的场所提供论坛(forum)。他的翻译思想是假设不存在共核和深层结构,将自己的解构理论建立在非等同、非对等和不可传达的基础之上。他认为,翻译是不断修改或推迟原文的过程以置换原文。他对翻译的定义是:翻译是一种语言对另一种语言、一种文本对另一种文本有调节的转换(翻译的性质)。德里达认为,各种语言在语义、句法和语音的差异上造成各不相同的表意方式,通过翻译我们对语言之间的差异性和各语言的特定表达方式可以达到更深刻和更准确的认识(翻译的作用)。因此,翻译的目的不是"求同",而是"存异"。一篇译文的价值取决于它对语言差异的反映程度和对这种差异强调的程度。德里达是从探索语言本质的角度来谈翻译的。他旨在强调通过翻译才能揭示语言之间的差异,而不是谈具体的翻译原则或方法。德里达用"转换"的概念取代"翻译"旨在说明,在翻译的过程中,通过修改和转换,原文在成长、成熟,最终得到"再生"(renewal)。

德里达还从解构主义的观点重新阐述了原文与译文的关系。他认为原文与译文的关系,不是传统翻译理论所主张的"模式—复制"的关系,而是一种"共生"的关系,即平等互补的关系。译文是另一个早先存在的译文的翻译,原先的译文又是更早的译文的翻译,如此向前不断循环,直至无限,即成为德里达所说的所谓"无限回归的意义链"。解构主义认为,一切文本都具有"互文性"(intertextuality),创作本身是一个无数形式的文本互相抄印翻版的无限循环的过程。"互文性"否定了原作的权威性与创造性,甚至连作者的著作权都不承认。这样,也就无所谓原文与译文之分了。这从另一个方面提高了译者和译文的地位。

(2) 韦努蒂的翻译策略:韦努蒂是解构主义翻译学派的另一代表人物。他详尽研究了自德莱顿以来的西方翻译史,批判了以往占主导地位的以目的语文化为归宿的倾向,提出了反对译文通顺的抵抗式解构主义翻译策略。他在《译者的隐身》中,对"通顺的翻译"提出质疑,韦努蒂认为,以往翻译传统是以民族中心主义和帝国主义的价值观来塑造外国文本的。其提倡的翻译原则就是"通顺的翻译"和"归化"的翻译。韦努蒂提出了反对译文通顺的翻译理论和实践。其目的不是在翻译中消除语言和文化的差异,而是要在翻译中表达这种语言上和文化上的差异。《译者的隐身》是指在译文中看不见译者的痕迹,即所谓"不可见性"(invisibility)。他提出了异化翻译策略,也是韦努蒂的"抵抗策略"(resistancy)。韦努蒂首先追溯了英美文化中归化翻译(domesticating translation)的历史。从10世纪的英国开始,通顺的翻译就成了英语翻译的规范。通顺的翻译要求译文读起来不像是翻译,而像英语原文创造的作品。这种主张,在理论上是把语言看成是交际的工具;在实践上强调通俗易懂,避免多义或歧义。在他看来,语言被看作是表达个人感情的手段,把翻译看作是恢复外国作家所要表达的意义;翻译的目的是使另一种文化变得可以理解。这一目的往往包含着一种把外国文本完全归化的危险。从这个角度来说,归化的翻译也起到了巩固目的语文化规范的作用。就英美文化而言,这是一种文化殖民主义表现。

韦努蒂写《译者的隐身》的目的,就是要反对传统的通顺策略,在译文中要看得见译者,以抵御和反对当今尤其是在英语国家中的翻译理论和翻译实践的规范。韦努蒂的翻译理论是以解构主义观点作为其理论基础的。他从解构主义的视角把翻译定义为:"翻译是译者在理解的前提下,用目的语的能指链来替代原语文本中的所指链的过程。"韦努蒂也追溯了异化翻译(foreignizing translation)的历史。异化的翻译保留了外国文本中之异,但破坏了目的语文化的规范,译文在忠于原文时,就背离目的语文化的规范。异化的翻译抑制民族中心主义对原文的篡改,在当今的世界形势下,尤其需要这种策略上的文化干预,以反对英语国家文化上的霸权主义,反对文化交流中的不平等现象。异化的翻译在英语里可以成为抵御民族中心主义和种族主义,反对文化上的自我欣赏和反对帝国主义的一种形式,以维护民主的地缘政治的关系。韦努蒂批评了当代英美的翻译流派中以奈达为代表的归化翻译理论。

在他看来,奈达是想把英语中透明话语的限制强加在每一种外国文化上,以符合目的语文化的规范。这是用通顺的翻译策略,把归化隐藏在透明度之中。这与其说是文化交流,还不如说是为了归化的目的对外国文本进行文化侵略。韦努蒂反对英美传统的归化,主张异化的翻译,其目的是要发展一种抵御以目的语文化价值观占主导地位的翻译理论和实践,以表现外国文本在语言和文化上的差异。

他称这种翻译策略为"抵抗"(resistancy):这种翻译不仅避免译文通顺,而且对目的语文化提出挑战,所谓"抵抗",就是抵抗目的语文化的种族中心主义。"抵抗式翻译"不适合于科技翻译,因为科技翻译主要是为了达到交际目的。

韦努蒂主张异化的翻译,是要发展一种理论的、批评的和文本的方法,并用这种方法把翻译作为研究和实践差异的场所,而不是当今普遍认为的那种同一性。

第二章 科技英语的文体特征与翻译

科技英语主要涉及科技著作、科研报告、科技论文、科普读物、实验方案与报告、生产企业的宣传材料和产品说明书等。这些文字资料所说明的内容都是客观的、准确的,在语言结构上有其独有的特征。根据科技英语的文体特征,译者在进行科技英语翻译活动时,应遵循相应的翻译原则,掌握相关的翻译技巧。本章主要介绍科技英语文体、科技英语翻译原则和翻译技巧。

第一节 科技英语文体

科技英语与文学语言等功能文体一样,具有共同的语音体系、语法体系和词汇体系,但是科技英语由于题材内容和使用方式的特殊性,与日常生活英语、文学题材的英语具有明显的区别。科技英语以其特殊的词汇、语法、语义和语用特征,以及在一般现在时、被动语态、无人称句、名词化、非动词化倾向和名词并列作定语的使用上,都体现出科技英语的独立文体特征。

一、科技英语的词汇特征

1. 专业性

专业性是指科技英语所使用的词汇具有专业特色,普通英语词汇专业化之后,成为具有特定含义和习惯用法的专业术语。例如,concrete 一词在普通英语中的意思是"具体的",但在土木工程中,concrete 的意思是"混凝土"。又如,current 在普通英语中意为"流行的,现行的",但在电力英语中,current 意为"电流"。

例1 Substance fall into three general groups: *conductors*, semiconductors and insulators.

物质分为三大类:导体、半导体和绝缘体。

conductor 在普通英语中的意思是"售票员,领导者,管理人"等。在该句中,conductor 的意思是"导体"。

2. 创新性

科技发展产生新事物,生成新概念,因而创造了新词汇或赋予原有普通词语新的科技含义。如计算机的发展产生了 absolute address(绝对地址),artificial intelligence(人工智能),CAD(计算机辅助设计),CPU(中央处理机),data(资料库),e-mail(电子邮件),firewall(防火墙),hardware(硬件),platform(操作平台),software(软件),terminal(终端)等新词;网络的发展产生了 internet(因特网,互联网),www(万维网)等新词;克隆技术的发展产生了 clone(克隆)等新词;太空技术的发展产生了 moonwalk(月球上行走),moonrock(月球标本石),space sickness(太空病),space age(太空时代),deep space(深层空间),black hole(太空黑洞)等新词。

例 2 The Shinkansen, also known as the bullet train, is a network of high-speed railway lines in Japan operated by four Japan Railways Group companies.

新干线,也就是人们熟知的子弹头火车,是由四家日本铁路集团公司运营的高速铁路网。

句中的 bullet train(子弹头火车)是高速铁路技术发展所产生的新术语。中国高铁的发展,汉语中产生了"动车组"一词,译成英语为:motor car unit。

3. 单一性

科技英语词汇的单一性特征是指在特定专业内,科技英语词语的词义具有单一性特征,主要体现在词语的专业性方面,同一个词语可能会在不同的专业中使用,成为多义词。但是,在专业人员看来,科技英语的多义现象其实并不成立,因为专业词汇或术语,一般都是经过严格的标准化定义,在特定专业内,词汇或术语的含义对专业人员来说往往是唯一的,具有不可替代性。

例 3 A contact is a conductive part in an electrical circuit attached to a switch that opens or closes a circuit by coming in contact with or separating from the main conductor.

接点是连接开关电路中的一个导电部件,通过与电源导线的接触或分离来打开或闭合电路。

单词 contact 具有多种含义,但在上句中,第一个 contact 作为电学继电器中的一个专业术语,其词义却是唯一的,即"接点"。

4. 精确性

科学技术的发展是理性思维或推理的产物,具有严格、严谨和严密等特征,科学技术文献的选词、词义、图表、数据等都要精确无误。所以,在科技英语翻译中,译者必须把握精确性特征,把理性思维的严谨、精确性,点点滴滴体现在译文的选词、组句、信息重组甚至译入语的表达习惯和词语搭配等方面。

例 4 The trains will literally float over the landscape levitated by powerful electromagnets cooled in liquid helium.

【译文1】这种火车借助液氦冷却的极强的电磁铁悬起,在田野上飘驰。("悬起"和"飘驰"均不准确)

【译文2】火车借助液氦冷却的强力电磁铁完全悬浮起来,在田野上飞驰。(较准确)

5. 客观性

科学的对象、方法和评价的客观性是科学客观性的三大内涵,决定了科技英语文献用词的客观性。同时,英语和汉语两种语言的差异性也是客观存在的,所以译者在翻译科技文献时,首先要以科学的态度,客观准确地选择科学术语或词语;其次还要将英汉两种语言结构不同、思维各异的语言统一起来,不能自由发挥、想当然,造成读者费解甚至是误解。

例 5 The temperature in the furnace is not always above 1,000℃.

【译文1】炉内的温度并不是每次都超过1 000℃。（不准确）

【译文2】炉膛内的温度并不总是处在1 000℃以上。（较准确）

二、科技英语的句法结构特征

1. 使用现在时

科技英语中最常见的动词时态是现在时态，包括一般现在时和现在完成时。因为一般现在时可以较好地说明科学概念、公理、定理、定义、公式的解说以及图表信息的精确无误，从而使内容更客观、信息更准确，提高科技内容的可信度。现在完成时用以表述已经发现的现象或取得的研究成果以及事物产生的影响。

例 6 A high voltage cable—also called HV cable—is used for electric power transmission at high voltage.

高压电缆，也称作HV电缆，用于输送高压电。

例 7 Machining requires attention to many details for a workpiece to meet the specifications set out in the engineering drawings or blueprints.

机械加工需要关注工件的许多细节，以满足工程图纸或蓝图上列出的技术要求。

2. 使用被动语态

语态是动词的一种形式，它表示主语和谓语的关系。如果主语是动作的承受者，或者动作不是由主语而是由其他人完成的，则用被动语态。由于科学所表现的客观性、可说明性、可验证性、抽象性等特征，作为传播科学的语言媒介，科技英语也具有客观性特征，叙述客观的事物，描述其发生、发展及演化的过程，阐明客观事物之间的内在联系，其主体通常是客观事物或自然现象，因此使用被动语态就显得更客观、更真实、更科学，并降低主观色彩。

例 8 Smaller generators are sometimes self-excited, which means the field coils are powered by the current produced by the generator itself. The field coils are connected in series or parallel with the armature winding.

较小的发电机有时是自励磁型的，即励磁线圈由发电机本身发出来的电流供电。励磁线圈与电枢统组串联或并联连接。

例 9 The magnetic field of the dynamo or alternator can be provided by either electromagnets or permanent magnets mounted on either the rotor or the stator.

通过安装在转子或定子上的电磁铁或永磁铁，可以形成直流发电机或交流发电机所需的磁场。

3. 使用无人称句

科技文体的主要目的在于阐述科学事实、科学发现、试验结果等，为了说明科学技术活动所带来的结果、证明的理论或科学发现，不过多地侧重人类的行为或发明者，因此，大多数科技文献都较少使用人称。

例 10 Most refrigerators, air conditioners, pumps and industrial machinery use AC power whereas most computers and digital equipment use DC power.

大多数冰箱、空调、泵和工业机械都使用交流电源，而大多数计算机和数字设备则使用直流电。

例11 The speed at which the rotor spins in combination with the number of generator poles determines the frequency of the alternating current produced by the generator.

转子的转速与发电机电极的数目共同决定发电机所发出的交流电的频率。

例12 Automation is the use of control systems and information technologies to reduce the need for human work in the production of goods and services.

自动化是利用控制系统和信息技术,减少商品生产和服务中对人力的需求。

4. 使用复杂的长句结构

为了将事理充分说清楚,科技英语中常常使用一些含有许多短语和分句的长句子,而短语和分句之间往往使用许多逻辑关联词,如 accordingly, consequently, finally, furthermore, however, then, therefore, in addition, on the contrary 等,使科技文献所表述的内容逻辑关系明确、结构层次分明、语言内容严密,更具有科学性。

例13 The availability of computers and microprocessors has completely changed the machine tool scenario by bringing in the flexibility which was not possible through conventional mechanisms.

计算机和微处理机的应用,通过引入传统机械系统无法实现的柔性生产方式,使机床得到彻底的变革。

例14 The development of Numerical Control in 1952 brought about a kind of flexibility to the metal cutting operation, therefore, at present a majority of manufacturing processes are making use of these principles in some form or the other, which allows for just in time manufacturing leading to zero inventories, zero setup times and single component batches without losing any advantages of mass manufacturing.

1952年,数控技术的发展在金属切削加工中产生了一种柔性生产方式。因此,目前大部分制造过程都在以某种方式使用这些原理。这就使零库存、零组装、实时生产成为可能,并且可以在不丧失批量生产优势的前提下实现单个零件的整批生产。

第二节 科技英语翻译原则

《现代汉语词典》(第 7 版)中对"原则"一词的释义为"说话或行事所依据的法则或标准"。从这一定义上来讲,翻译原则是对整个翻译过程进行导向,而不是作为评判翻译结果的标准。基于上面的表述,"原则"与"标准"是完全不同的概念,不能等同视之。否则,会对翻译理论的研究造成极大妨碍。翻译原则包含四项内容:前后一致、直译优先、约定俗成和灵活有度。这四项翻译原则是译者在整个翻译过程中必须坚守的译事法则,为整个翻译过程确立了基本原则,对生产合格的翻译产品具有重要意义。翻译作为"产品"必须按照翻译界通行的适用翻译标准进行评价,这就是翻译标准与翻译原则之间的差别。

一、前后一致

所谓前后一致,是指译文前后逻辑要一致,切不可"驴唇不对马嘴"。就科技英语翻译而言,要求术语或词语翻译必须前后一致,以免引起误解,这也是翻译的基本原则。例如,"paste backfill"翻译成汉语是"膏体充填料",在整篇译文里要求一律采用同一译法,绝不可又将其译为"牙膏充填料"。这就是"前后一致"的体现。

一些译者不了解这一翻译原则,使得同一个词语出现多种译法,造成前后矛盾,弄得读者一头雾水,进而影响到整个翻译的效果。

例1 The sedimentary rocks associated with coal beds belong to one or more of the four classes of compacted strata given in Table 2. When certain sediments were deposited to form limestone or *sandstone*, we could hardly expect them to consist of pure lime deposits or pure sand; there would be some mixing of *sediments*. As a result, we find limestone containing small amounts of sand, and sandstone containing small amounts of lime carbonate, etc. A rock belonging to any one of four main classes which, in addition, contains sand is classed as a sandy stone; when small amounts of lime are present, the name limey is applied; with small amounts of silt and clay, the name shaly is applied.

与煤层伴生的沉积岩属于表2所列的四种致密地层中的一种或几种。当某些沉积物沉积成石灰岩或砂岩时,我们很难想到它们是由纯的石灰或砂的沉积物所组成的,其中一定会有某些沉积物的混合物。因此,我们会发现石灰岩里含有少量砂,而砂岩里含少量碳酸钙等。此外,凡属于上述四种主要分类中的任何一种岩石,含砂的则划为砂质岩一类,存在少量石灰的称为钙质岩,而含有少量粉砂和黏土的则称为泥质岩。

例1中斜体的词都是在这段短文中重复出现的,翻译时,应注意前后一致,不可有多种译法,这点要切记。

除个别情况(因为英语有一词多义现象)之外,翻译时应保持同一词语同一译法,以免通篇译文出现前后矛盾的现象。需要指出的是,不论选词还是源语理解,译者的思维一定要跟着源语的话题走,这样就很少会出现逻辑思维上的错误。

二、直译优先

直译,用朱光潜教授的话说,就是依照原文的字面翻译,逐字逐句地翻译,字间的顺序也不能更改。但是,在实际翻译中,用直译的方法往往解决不了问题,必须结合意译。如此看来,直译与意译如同一对孪生姊妹,究竟应该怎样处理它们之间的关系呢? 1979年,董乐山先生在《翻译通讯》第2期中撰文指出,"能够尽量做到概念与字面都对等当然最好,在两者不能兼顾的情况下,为了忠实表达原意,就需要取概念而舍字面,这也就是说,要传达内容而不拘泥于形式"。就科技英语翻译而言,首先讲求"信",而直译是实现"信"的最佳翻译方法。因此,在翻译过程中通常优先考虑直译,这是译者应遵循的翻译原则。

例2 Paste backfill is a high density mixture.

膏体充填料是一种高密度混合料。

这句话用直译就可在内容和形式上达到"信"的效果,而不必采用意译,这就是"直译优先"原则。但是,并不是所有的源语都能采用直译,如下面这种情况只能采用意译:

例3 The cost of operating a pump over time is where most companies feel the impact.

大部分公司觉得有影响的是泵累积运转的费用。

例 3 中的句子就不能一词一词地翻译,应采用意译,以求内容的忠实而不求形式的一致。在此需要指出的是,运用意译时,要求译者对原文的理解有十足的把握。

科技英语的专业性较强,如果译者对源语涉及的专业不熟悉,在不能直接采用直译的情况下,最好和相关专业人员沟通,用意译的方法把源语的内容完整地翻译出来。这是不得已的办法。保持源语在翻译过程中不"走调",就是笔者强调坚持"直译优先"原则的理由。

三、约定俗成

所谓约定俗成,是指在翻译过程中碰到某些已经有固定译法的词语,译者不可再用其他译法,应直接采用其固定译法进行翻译,以免读者发生误解。例如,New York 这类词,已有"纽约"这种固定译法,就不能译成"新约克",这就是约定俗成原则。

在科技英语中,很多词语都有固定译法,这就需要译者日常不断积累。值得指出的是,同一词语可能有多种固定译法,这就要求译者甄别后选取最合适的译法。如果不坚持约定俗成原则,就有可能出现翻译错误的问题。例如,一份工程施工报告中,把 Porland cement(波特兰水泥)译成"港口陆地水泥"。就日常英语而言,固定译法的词语不胜枚举。比如"中华人民共和国",它的译文是固定的,即 the People's Republicof China,决不可有第二种译法。

按照约定俗成原则翻译的词语类型主要有人名、地名、企事业单位名称、专业术语、行话等。若译者在翻译过程中碰到这类词语,不能确定如何翻译时,一是查词典,二是上网查询,三是多请教他人,不能想当然地翻译。

四、灵活有度

有些译者在翻译过程中,总喜欢即兴发挥,有意将词义引申或添加译者的理解。这种做法称不上是灵活翻译,而是一种"强加于人"的译法,有悖于对原文的忠实性。

所谓灵活有度,意思是说,在求"信"的基础上,尽量灵活翻译不要"死译"或"硬译"。这就要求译者对准确理解源语有十足的把握,否则会弄巧成拙。

求"信"是灵活翻译所要掌握的"度"。如果翻译灵活、行文如水,而内容却离题万里,这种翻译也不是"灵活有度"。就科技英语翻译而言,有些理工科专业的译者容易犯这样的错误,他们凭借自己的技术特长,在翻译过程中随意发挥,一般不对照原文,不容易看出翻译的不妥之处;而有些外语专业的译者却"胆子"较小,常犯"硬译"的错误。

例 4 The slump test is related to the yield stress due to the fact that concrete slumps or moves only if the yield stress is exceeded and stops when the stress is below the yield stress.

塌落试验与屈服应力有关,因为只有当屈服应力过高时,混凝土才会塌落或移动,但在混凝土应力低于屈服应力时,混凝土会静止不动。

例 4 是科技英语翻译培训班出的一道课外作业题,有的译者翻译得灵活有度,并不拘泥于原文的形式,而是讲求内容的忠实性,灵活地把 due to the fact that 翻译成"因为",如果按照源语的句子结构或词序翻译,就会使译文生硬、洋化。

一般刚涉足翻译领域的人,不太容易做到灵活有度,因为灵活的分寸很难掌握。解决这一问题的有效途径就是多进行翻译实践。要想在翻译过程中把握好灵活性的分寸,靠的是译者对源语的领悟力。倘若译者不具备这样的条件，还是直译为好。

在科技英语翻译中，要坚持灵活有度这一翻译原则。科技英语翻译讲求科学严谨，所以给译者灵活发挥的机会不多，这就要求译者把握好翻译的灵活性，不要轻易阐译、释译或意译。

第三节 科技英语翻译技巧

在科技英语翻译中，常用的翻译技巧包括直译法、转译法、增译法、省译法、分译法、合译法等。通常，针对不同的翻译对象以及不同的文本类型，可以选用不同的翻译技巧以及这几种方法的结合。

一、直译法

纽马克认为，直译是在将源语的主要含义翻译成目的语时，尽管语言环境变化，也要尊重源语的句法结构的翻译方法。在转换过程中，直译尽可能地力图保持源语的形式结构，如词序、句子结构、修辞效果等，同时力求译文做到忠实、通顺、容易理解。

运用直译必须掌握一定分寸，直译不是死译、硬译，不是"逐词翻译"。纽马克指出的"逐词翻译"是将原文的语法结构以及所有词汇的意思一词不漏地严格翻译出来的方法，这种方法仅适用于翻译一些非常简单的句子。这与本节要讨论的"直译"是有很大区别的。

运用直译法的原因主要基于：

（1）源语与目标语的对等性和相似性。源语的基础词汇以及语法关系通常能在目标语中找到相同或相似的表达，这就是直译法赖以生存的基础所在。

（2）翻译家严复提出"信、达、雅"的翻译标准中的"信"即"忠实"翻译，既包括对原文内容的忠实，也包括对作者的忠实，从广义上讲，甚至还包括对译文目标读者的忠实。按照这个标准去考查，只有直译才能达到这样的效果。

一般来说，如果英语原文句型与汉语的句法规律较接近，词序也形似，意思又比较明白，那么可以用直译。这样就可以较便捷地得到意思准确、文字通顺的译文。

例1 A step-down transformer can reduce voltage to whatever is desired
降压变压器能把电压降低到所需要的任何数值。

例2 The igniter combustion often produces hot condensed particles.
这种点剂燃烧常常产生生热的凝结颗粒。

例1中的译文主干部分与原文基本一致，所以采用了直译法，既保持了原文的结构，又正确表达了原文的内容。例2译文通过直译保留了原文的词序和句子结构，使得内容表达通顺。

二、转译法

在翻译过程中，由于英汉两种语言在语法和表达习惯上的差异，要想将每个英语的词汇都翻译成和其词义以及词性一样的汉语词是很难实现的，有时必须改变原来某些词语的词类或句子成分才能有效地传达原文的准确意思。科技英语虽然没有文学英语那样在翻译上有较大的自由度,但要想使译文能符合汉语的语法和修辞习惯,在很多情况下都要作一些"修饰",如语序的重大调整、句子成分的改变、句子结构和句子表达方式的转换、词义的引申等。转译法指的是在进行翻译的过程中,为了实现有效、准确地表明原文的意思而把相关词汇的词义或词性进行改变。

1. 英译汉词义的转换

翻译时,如果完全生搬硬套词典所给的字面意思来"对号入座",会使译文生硬晦涩,含糊不清,甚至令人不知所云。因此,遇到一些无法直译的词或词组时,应根据上下文和原词的字面意思,作适当的转化,对其意义加以引申,但不能改变原意,也不改变其词性和句子成分。

例1 Solar energy seems to *offer more hope* than any other source of energy.

太阳能似乎比其他能源更有前途。

例1中的 offer more hope 本义为"提供更多希望",此处转译为"更有前途"。

2. 英译汉词性的转换

英汉两种语言中词类不尽相同,各词类的使用频率、使用方式也不尽相同,因此,原来英语中属于某种词类的词,在译成汉语时可以转为或必须转为另一种不同的词类。具体来讲,有些词类英语中有而汉语中没有,如冠词、引导词、关系代词、名词性物主代词、关系副词、分词、动词不定式、动名词等等。有些词类在汉语中有而在英语词汇中却没有,如助词。词类划分的差异必然导致词类转换法的广泛使用。

(1) 由于英语简单句只有一个动词,汉语句子却有多个动词,英语中的名、形容词、副词、动名词和介词都可以转译为动词。

例2 If extremely low-cost power ever to become *available* from large nuclear power plants, electrolytic hydrogen would become competitive.

如果能从大型核电站获得成本极低的电力,电解氢就会有更强的竞争力。

例3 In this case the temperature in furnace is *up*.

在这种情况下,炉温就升高。

例2中,形容词 available 转译为动词"获得"。例3中,形容词 up 转译为动词"升高"。

例4 Proper *selection* of circuit components permits a transistor to operate in this characteristic region.

正确地选择电路元件能使晶体管在这个特征区工作。

例5 *Reversing* the direction of the current reverses the direction of its lines of force.

倒转电流的方向也就倒转了其磁力线的方向。

例4中,名词 selection 转译为动词"选择"。例5中,动名词 Reversing 转译为动词"倒转"。

(2) 英语的动词、代词、副词及形容词均可根据具体情况转译为汉语的名词。

例6 Such materials are *characterized* by good insulation and high resistance to wear.

这些材料的特点是绝缘性好和耐磨性强。

例6中,动词 characterized 转译为名词"特点"。

英语中有些动词在翻译成汉语时很难找到相应的动词,这时可将其转译成汉语名词。

例7 An electric current *varies* directly as the electromotive force and inversely as the resistance.

电流的变化与电动势成正比,与电阻成反比。

例8 Boiling point is *defined* as the temperature at which the vapor pressure is equal to that of the atmosphere.

沸点的定义就是蒸汽的气压等于大气压时的温度。

例7中的动词varies(directly, inversely)转换成名词"(正)比,(反)比"。例8中的动词过去分词defined转换成名词"定义"。

英译汉中所谓代词转译为名词实际上就是把代词所代替的名词翻译出来。

例9 The radioactivity of the new element is several million times stronger than *that* of uranium.

这种新元素的放射性比铀的放射性要强几百万倍。

例10 Radio waves are similar to light waves except that *their* wave length is much greater.

无线电波与光波相似,只不过无线电波的波长要长很多。

例9中的代词that转译为其所指代的名词"放射性"。例10中的their转译为其所指代的名词"无线电波"。

英语中有些副词在句子中用作状语,译成汉语时可根据具体情况转换成汉语的名词。

例11 Such magnetism, because it is *electrically* produced, is called electromagnetism.

由于这种磁性产生于电,所以称为电磁。

例12 Oxygen is one of the important elements in the physical world, and it is very actively *chemically*.

氧是物质世界的重要元素之一,其化学性能很活泼。

例11中的副词electrically转译为名词"电"。例12中的副词chemically转译为名词"化学"。

英语中有些形容词加上定冠词the表示某一类人或物,汉译时常译成名词。另外,英语中有些表示事物特征的形容词作表语时,往往也可在其后加上"性""度""体"等词,译成名词。

例13 In fission processes the fission fragments are very *radioactive*.

在裂变过程中,裂变碎片的放射性很强。

例14 The more carbon the steel contains, *the harder and stronger* it is.

钢含碳量越高,强度和硬度就越大。

例15 The *electrolytic* process for producing hydrogen is not so efficient as the *thermochemical* process.

用电解法生产氢气的效率不像热化学法那样高。

例13中的形容词radioactive转换成名词"放射性"。例14中的形容词the harder and stronger转换成名词"强度和硬度"。例15中的形容词electrolytic, thermochemical分别译成名词"电解法""化学法"。

(3) 英语的名词、副词及动词可译成汉语的形容词。

例16 Gene mutation is of great importance in breeding new *varieties*.

在新品种的培育方面,基因突变是非常重要的。

原句中的 breeding new varieties 中的 varieties 转译为形容词"新品种的"。

当英语动词转译为汉语的名词时,修饰该动词的副词往往也随之转译为汉语的形容词。

例17 This man-machine system is *chiefly characterized* by its simplicity of operation and the ease with which it can be maintained.

这种人机系统的主要特点是操作简单,容易维修。

例18 The equations *below* are derived from those *above*.

下面的方程式是由上面的那些方程式推导出来的。

例17中的动词过去分词 characterized 转译为名词"特点",其修饰副词 chiefly 相应转译为形容词"主要"。例18中的副词 below,above 分别转译为形容词"下面的""上面的"。

例19 Light waves *differ* in frequency just as sound waves do.

同声波一样,光波也有不同的频率。

原文的动词 differ 转译为形容词"不同的"。

(4) 英语中能转译为副词的主要是形容词,有时动词也可转译为副词。

当英语的名词转译为汉语的动词时,原来修饰名词的英语形容词就相应地转译为汉语的副词。

例20 In case of use without conditioning the electrode, *frequent calibrations* are required.

如果在使用前没有调节电极,则需要经常校定。

原文的名词 calibrations 转译为动词"校定",修饰 calibrations 的形容词 frequent 转译为副词"经常"。

在系动词加表语的结构中,作表语的名词转译为汉语的形容词时,原来修饰名词的英语形容词就相应地转译为汉语的副词。

例21 This experiment is an absolute *necessity* in determining the solubility.

对确定溶解度来说,这种试验是绝对必要的。

原文中作表语的名词 necessity 转译为汉语的形容词"必要的",修饰它的形容词转译为汉语的副词"绝对"。

例22 In actual tests this point is *difficult* to obtain.

在实际的测试中,很难测到这个点。

例23 The mechanical automation makes for a *tremendous* rise in labor productivity.

机械自动化可以极大地提高劳动生产率。

例22中的形容词 difficult 转译为副词"很难"。例23中的形容词 tremendous 转译成副词"极大地"。

当英语中的谓语动词后面的不定式短语或分词转译为汉语的谓语动词时,原来的谓语动词就相应地转译为汉语的副词。

例24 Rapid evaporation *tends to make the steam wet*.

快速蒸发往往使蒸汽的湿度加大。

原文的不定式短语 to make the steam wet 转译为谓语动词"使蒸汽的湿度加大",动词 tend 转译为副词"往往"。

(5) 英语的介词可译成汉语的动词。

例25 Figure 2.3 gives these results *in* the form of a frequency distribution or histogram.

图 2.3 用频率分布或频率曲线来表示这些结果。

例 25 中的介词 in 转译为动词"用",像这样的例子还很多。

3. 英译汉句子成分的转换

转译是英汉翻译中常采用的一种调整、变通甚至是创造性的翻译技巧。在翻译中,不仅词类可以转译,句子甚至是段落和篇章都可以转译。

一个或多个词的词性发生了变化,其语法功能势必也会随之发生变化,那么该词在译文中的成分和功能也会相应变化。例如,如果翻译中将动词转译为名词,那么修饰该动词的副词也会随之转译为形容词,转译后的名词可能成为主语。接下来主要探讨英译汉中主语的转译、谓语的转译、宾语的转译、表语的转译、定语的转译、状语的转译和补足语的转译。

(1) 主语的转译。

例26 Both *the frequency and inductance* of a crystal unit are specified for inclusion in a particular filter.

晶体单元要包含在一个特定的滤波器中,就要规定出频率和电感。

例27 *The mechanical energy* can be changed back into electrical energy by means of a generator.

利用发电机可以把机械能重新转变成电能。

例28 A *semiconductor* has a poor conductivity at room temperature, but it may become a good conductor at high temperature.

在室温下半导体的电导率差,但在高温下它可能成为良导体。

例 26 中的主语 both the frequency and inductance 转译为宾语"频率和电感"。例 27 中的主语 the mechanical energy 转译成介词宾语"机械能"。例 28 中,主语 semiconductor 转译为定语"半导体的"。

(2) 谓语的转译。

英语句子中的谓语一般情况下不需要作特殊处理。但是,有些谓语动词很难直接翻译成符合汉语习惯的动词,这时就需要把其进行适当转换,其他成分也相应跟着转换。

例29 These pumps *are featured* by their simple operation, easy maintenance, low consumption and durable service.

这些水泵的特点是操作简便、维修容易、耗油量少、经久耐用。

例30 The sun *produces* in three days more heat than all earth fuels could ever produce.

太阳在三天内发出的热量就比地球上所有的燃料发出的热量还要多。

例31 Water with salt *conducts* electricity very well.

盐水的导电性能良好。

例 29 中的谓语 are featured 转译为主语"特点"。例 30 中,谓语 produces 转译为定语

"发出的",宾语 heat 相应转译为主语。例 31 中,谓语 conducts 转译为定语"导电(的)"。

(3) 宾语的转译。

英语句子的宾语,在翻译时可以转译为汉语的主语、谓语、定语等成分。

例32 The production had considerable *difficulty* getting patent protection if it had no patent right.

如果产品没有专利权,要获得专利保护是相当困难的。

例33 A motor is similar to a generator in *construction*.

电动机的结构跟发电机类似。

例34 Light beams can carry more information than *radio signals*.

光束运载的信息比无线电信号运载的信息多。

例 32 中的宾语 difficulty 转译成表语"(是)相当困难的"。例 33 中的介词宾语 construction 转译为主语"结构",注意此时介词不译。例 34 中的介词宾语 radio signals 转译为定语"无线电信号运载的"。

(4) 表语的转译。

英语中系表结构很多,翻译时往往需要转译为其他成分。

例35 Rubber is *a better dielectric* but a poorer insulator than air.

橡胶的介电性比空气好,但绝缘性比空气差。

例36 Another possibility of using solar energy is *in house-heating*.

住宅供暖是太阳能的另一种可能用途。

例37 Also present in solids are *numbers of free electrons*.

固体中也存在着大量的自由电子。

例 35 中的表语 a better dielectric 转译成主语"介电性"。例 36 中的表语 in house heating 转译为汉语中的主语"住宅供暖"。例 37 中的表语 numbers of free electrons 转译成宾语"大量的自由电子"。

(5) 定语的转译。

英语中表示性质的定语,比如形容词或分词,往往暗含谓语的意味,常常会转译为谓语动词。此外,如果英语中某一名词转译成了汉语动词,那么修饰该名词的形容词或分词定语也要相应地转译为汉语的状语。

例38 The largest power stations, *being the most efficient*, are run all the time at full load.

一些最大的发电厂,因为它们效率高,总是满负荷运行。

例39 A semiconductor has a *poor* conductivity at room temperature, but it may become a good conductor at high temperature.

在室温下半导体导电率差,但在高温下它可能成为良导体。

例40 Copper, which is used so widely for carrying electricity, offers very *little* resistance.

铜的电阻很小,因此我们广泛地用铜来输电。

例 38 中的分词短语 being the most efficient 是主语的定语,但它具有状语意义,表示原因,所以转译成原因状语。例 39 中的定语 poor(conductivity)转译成谓语"(导电率)差"

例 40 中的 little(resistance)转译成谓语"(电阻)很小"。

(6) 状语的转译。

英语中有一些介词短语,常常在意义上和主语有密切的关系,表示主语的位置、状态和性质等,往往会转译为汉语中的主语。此外,如果被修饰词的词性发生变化,如英语的副词转译为汉语中的定语时,那么,一些作状语的介词短语常常转译为定语。

例41 The wide applications of electricity affect *tremendously* the development of science and technology.

电的广泛应用,对科学技术的影响极大。

例42 This communication system is chiefly characterized by *its simple operation and easy maintenance*.

操作简单、维护方便是这个通信系统的主要特点。

例43 The attractive force between the molecules is *negligibly* small.

分子间的吸引力小得可以忽略不计。

例 41 中的状语 tremendously 转译成谓语"极大"。例 42 中的 by its simple operation and easy maintenance 转译成主语"操作简单、维护方便"。例 43 中的状语 negligibly 转译为汉语中的补语"可以忽略不计"。

(7) 补足语的转译。

英语中补足语分为主语补足语和宾语补足语,可转译为汉语中的谓语和宾语。

例44 The flow of electrical charges *makes the coil* a magnet.

电荷的流动使线圈成为一块磁铁。

例45 Water resources are reported *to have great importance for hydroelectric power stations*.

据报道,水资源对于水电站具有重大意义。

例 44 中的宾语补足语转换为宾语。本句的成分转换比较复杂,译文中加"成为"二字,使译句正确、通顺。"成为一块磁铁"为汉语兼语句中的第二谓语部分,其中"一块磁铁"为"成为"的宾语。例 45 中的补足语 to have great importance for hydroelectric power stations 转译为谓语"对于水电站具有重大意义"。

有时句子成分的转换具有连锁反应,一个成分变了,其他成分也要随之改变,具体情况需要具体对待。但必须明确,进行成分转译的目的是为了使译文通顺,合乎汉语习惯和更好地与上下文保持一致。

4. 英译汉句子结构的转换

在翻译一个句子时,虽然每个单词或词组似乎都理解对了,但译文在整体上不利落或不地道。这是因为英汉两种语言之间的差异不仅体现在词汇层面上,而且体现在句子层面上,二者在句子结构上还存在很大的差别。这主要表现为:英语主语突出,汉语主题突出;英语组句多焦点透视,句式呈树式结构,汉语组句多散点透视,句式呈竹式结构;英语思维重逻辑,句式严谨规范,缺少弹性,汉语思维重语感,句式长长短短,灵活多变。

英译汉时,由于两种语言的句子结构大不相同而往往需要改变一下句子结构以适应于汉语的表达习惯,包括把英语简单句译成汉语复句,或英语复合句译成一个汉语单句,或一种从句转译成另一种从句,比如把状语从句转译成条件偏正复句等。

(1) 英语简单句译为汉语复句。

例46 Vibrating objects produce sound waves, each vibration producing one sound wave.

振动着的物体产生声波,每一次振动产生一个声波。

此句英语简单句译成汉语联合复句。

(2) 英语复合句译成汉语简单句。

例47 Good clocks have pendulums which are automatically compensated for temperature changes.

优质的钟摆可以自动补偿温度变化造成的误差。

例48 The turbines drive the dynamos which generate the electricity.

涡轮机带动发电机发电。

例47中含有定语从句的英语复合句转译为汉语简单句。例48中同样把含定语从句的复合句转换为了汉语简单句。

(3) 复合句变成复句。

例49 Where there is nothing in the path of beam of light, nothing is seen.

假如光束通道上没有东西,就什么也看不到。

例50 The molecules exert forces upon each other, which depend upon the distance between them.

分子相互间都存在着力的作用,该力的大小取决于它们之间的距离。

例49中含地点状语从句的英语复合句转译成汉语条件偏正复句。例50中含定语从句的英语复合句译成汉语的联合复句。

5. 英译汉表达方式的转换

与汉语比较,英语科技文体的一个突出特点是"主观成分少,着重客观叙述",因此多用被动形式。科技文章广泛使用被动语态的第二个主要原因是将主要信息前置,放在主语位置。相反,汉语则多采用主动形式来表达。此外,英语中的肯定(否定)语气也可以转换成汉语的否定(肯定)语气。

(1) 语态的转变。

在将科技文章译成汉语的过程中,需将被动句转换成主动句、自动句或无主句,以符合汉语的表达习惯。当英语被动句中的主语为无生命的名词,又不出现由介词 by 引导的行为主体时,往往可译成汉语的主动句,原文的主语在译文中仍为主语。这种把被动语态直接译成主动语态的句子,实际是省略了"被"字的被动句。

例51 With this information *the " phase" of the coherent train of pulses* could be traced in addition to simply determining the pulse rate.

有了这一信息,除能确定脉冲速率外,还可寻找出这一连串脉冲的相位。

例52 After *sealing the header* is cleaned and then *the leads* are clipped to the desired length.

封焊后把管座清洗干净,然后把引线剪到所需长度。

例53 Friction can be reduced and the life of the machine prolonged *by lubrication*.

润滑能减少摩擦,延长机器寿命。

例51中的原文主语the "phase" of the coherent train of pulses译成宾语"脉冲速率"。例52中,把原文主语译成宾语,而把行为主体或相当于行为主体的介词宾语译成主语。含有by的英语被动句,有些也可以转译成汉语的主动语态。其中介词by后的宾语转译为主动语态中的主语,按照这个主语的人称和数以及原来的时态把谓语动词形式由被动语态改为主动语态;或者by短语直译为状语,原文主语转译为谓语,句子译为无主句。例53中,by的宾语lubrication转译为主语"润滑",整个原文由被动语态转为主动语态。

(2) 语气的转变。

首先,否定语气译为肯定语气。

例54 It makes *no difference* whether such a shelter is built of stones or wood.

这种棚子用石头砌还是用木头建都一样。

其次,肯定语气转译为否定语气。

例55 Although the power plant has a long history, most of the equipments *are relatively new*.

这个电厂虽然历史悠久,但大部分设备并不老旧。

例56 The experiment is *far from being satisfactory*.

试验结果远不能令人满意。

例54中的makes no difference转译成肯定语气"一样"。例55中的are relatively new转译成否定语气"不老旧"。例56中的far from being satisfactory含有否定意义,译文采用否定形式"远不能令人满意"。

三、增译法

英汉两种语言有不同的表达方式。在翻译过程中,要对语意进行必要的增减。在句子结构不完善、句子含义不明确或词汇概念不清晰时,需要对语意加以补足,从而在语言形式或者表达习惯上更加符合译文的文化背景或者语言习惯。

增译法是翻译时按意义、修辞和句法上的需要把英语中没有的语言成分加进译文中,从而使译文能确切地表达原文的内容,符合汉语语言的表达习惯。增译的原则是增加那些在语义上或修辞上必不可少的词语,也就是增加原文字面虽无其词而有其意的一些词。

为了使行文简洁紧凑,避免不必要的重复,英语经常将句中某些成分省略。但是汉译时必须予以增补,否则就会使译文语言不畅、语法不通或语义不明。

例1 With the development of modern electrical engineering, power can be transmitted to wherever it is needed.

随着现代化电气工程的发展,人们可以把电力输送到任何所需要的地方。

例1中增加"人们",原文由被动语态转为主动语态。

另外,许多英语句子中虽然并未省略某些成分,但翻译时仍然必须增补有关词语才能使汉语译文明快达意,文从字顺。

(1) 增译名词。

英语中由动词或形容词派生而来的抽象名词,翻译时可根据上下文在其后面增译范畴词,如"部分""过程""状态""现象""工作""方法""情况""作用""装置""效应"等,使抽象名词具体化,更符合汉语表达习惯。

例2 Due care must be taken to ensure that the pulse signal itself shall show *no irregularities and no interruptions*.

应注意保证脉冲信号本身不出现不规则现象和中断现象。

例3 The cost of such a power plant is a relatively small portion of the total cost of the development.

这样一个发电站的修建费用仅占该开发工程总费用的一小部分。

例4 To get an idea of electrical pressure, let us consider a simple analogy.

为了得出电压的概念,让我们考虑一个简单的类似情况。

例2中的irregularities与interruptions分别由形容词和动词派生而来,根据上下文增译"现象"。例3中的development后增译了"工程"。例4中的analogy后增译了"情况"。

(2) 增译副词。

英语的时态是通过动词的曲折变化来实现的,而汉语的时态是增加表示时态的助词或表示时间的副词来完成的。因此,在翻译英语的进行时态时常常增译"正在""在""不断""着"等等,将来时态时需增译"将""要""便""会"等等;过去时态时需增译"曾经""当时""以前""过去"等等;完成时态时增译"已经""历来"等等。

例5 The shell parts of reactor pressure vessels have been fabricated with formed plates welded together.

反应堆压力容器的壳体,历来是用成型钢板焊接而成的。

例6 The solution is about to boil.

溶液马上就要沸腾了。

例5增译副词"历来"表示完成时态。例6增加副词"要"表示将来时态。

(3) 增译概括性的词语。

所谓概括词语就是类似于"两种""三类""双方""等等""种种"等等。增译概括性的词就是将所罗列的事物用概括性的词来总结概括。

英语和汉语都有概括词。但由于语言习惯上的差异,两者使用这些词语的场合却不尽相同。因此,有时英语原文中并无概括性的词语,但译成汉语时却需要增补"两人""双方""等""等等""凡此种种""几方面""各种"等概括词,才能使译文更加符合汉语的修辞规范。

例7 For reasons the alternating current is more widely used than the direct current.

由于种种原因,交流电比直流电用得更为广泛。

例8 The frequency, wave length and speed of sound are closely related.

频率、波长和声速三者是密切相关的。

例9 This report summed up the new achievements made in electron tubes, semiconductors and components.

本报告总结了电子管、半导体和元件三方面的新成就。

例10 Sensor switches are located near the end of each feed belt.

各传感器开关位于每条给料皮带的末端附近。

例11 A data processor can issue address and function codes.

数据处理器能发出各种地址码和功能码。

上述例句中分别增译了概括性词语"种种""三者""三方面"和不可忽视的复数表达"各"和"各种"。

（4）增译连接词。

为了使译文更加富有逻辑，更加符合汉语的表达习惯，更加通顺流畅，可以适当地在译文中增加表示原因、条件、目的、结果、让步、假设等的连接词。

例12 As the nature of the soil often varies considerably on the same construction site, the capacity of the soil to support loads also varies.

即便在同一施工工地，由于地基的性质有很大差异，土地的承载力也不相同。

此外，在英语中常用倒装语序表示的虚拟条件状语从句，译成汉语时，往往可增加"如果……那么……""假如……就……""万一……就……"等连接词。

例13 Should a DC system be used, the losses in transmission would be very great.

如果使用直流系统的话，输电线路上的损耗就会很大。

例14 Were there no electric pressure in a conductor, the electron flow would not take place in it.

导体内如果没有电压，便不会产生电子流动现象。

例13中的倒装句中增译了连接词"如果……就……"。例14中的倒装句翻译为正常语序，增译了连接词"如果……便……"。

以不变化的be开头的让步状语从句，在做英译汉时，往往增加"不论""不管""无论"等连词。

例15 All magnets behave the same, be they large or small.

所有磁体，无论大小，其性质都一样。

如果分词短语或独立分词结构含有时间、原因、条件、让步等状语意义，翻译时可增加"当……时""……之后""因为……""由于……""如果……""虽然……但是……"等词。

例16 Using a transformer, power at low voltage can be transformed into power at high voltage.

如果使用变压器，低压电就能转换成高压电。

例16中分词短语作条件状语，所以增译连接词"如果……"。

表示目的或结果状语的动词不定式和不定式短语通常在其前面加"为了""要""以便""结果""就""从而"等词。

例17 Crimp hinge ends after pin installation to retain pin.

销子安装后，将铰链两端紧固，以免销子脱落。

例18 We made transistors by different means only to get the same effect.

我们用不同的方法制造晶体管，结果得到相同的效应。

例19 Many lathes are equipped with multistage speed gearboxes to get different speeds.

为了得到不同的速度，许多车床装有多级变速齿轮箱。

例17增译了连接词"以免"。例18的不定式短语only to get the same effect表示结果，增译连接词"结果"。例19中的不定式短语to get different speeds表示目的，增译连接词"为了"。

四、省译法

省译法是指在翻译的过程中在不改变原意的基础上省略掉一些不必要的成分。这种做法,一般是出于译文语法或者表达习惯的需要。

(1) 英译汉时可省译代词。

例1 He has distinguished *himself* in power plant construction.

他在电厂建设方面表现出色。

例2 A wire lengthens while *it* is heated.

金属丝受热则伸长。

例1省译了主语的反身代词himself,例2省译了形式主语的代词it。

例3 Electrical leakage will cause a fire, hence you must take good care of *it*.

漏电会引起火灾,因此必须多加注意。

例4 Friction always manifests *itself* as a force that opposes motion.

摩擦总是表现为一种对抗运动的力。

上面两个例子中的himself、代词it及作宾语的反身代词itself均省译。

英语中的物主代词,汉译时往往也可以省译。

例5 Different machines differ in *their* mechanical properties.

不同的设备具有不同的机械性能。

例6 The diameter and the length of the wire are not the only factors to influence *its* resistance.

导线的直径和长度不是影响电阻的唯一因素。

例5省译了物主代词their,例6省译了物主代词its。

(2) 英译汉时可省译连接词。

汉语的逻辑关系常常是暗含的,由词语的次序来表示。英语则多用连接词。因此,在英译汉时,很多情况下不必把连接词译出来。

例7 Like charges repel each other *while* opposite charges attract.

同性电荷相斥,异性电荷相吸。

例8 The advantage of rolling bearing is *that* they cause less friction.

滚动轴承的优点是它产生的摩擦力较小。

例9 The volume of a given weight of gas varies directly with the absolute temperature, *provided* the pressure does not change.

压力不变,一定重量的气体的体积与绝对温度成正比。

上述三例中的并列连接词while、连词that及条件连词provided均省译。

(3) 英译汉时可省译介词。

英语中介词使用频率高,用来表示名词或代词等与句中其他成分的关系。相对而言,汉语中的介词较少。英译汉时,许多介词可省略不译。

首先,省译某些名词、形容词或动词后面搭配的介词。

例10 When we talk of electric current, we mean electrons *in* motion.

当我们谈到电流时,我们指的是运动的电子。

例11 Whenever a current flows through a resistance, a potential difference exists *at* the two ends of the resistance.

电流通过电阻时，电阻的两端就有电势差。

例10中的介词in被省略，例11中的介词at被省略。

其次，省译某些介词短语中的介词。

例12 Sounds having the same frequency are *in resonance*.

具有相同频率的声音会共振。

例13 *In the transmission of* electric power a high voltage is necessary.

远距离输电必须用高压。

例12中的介词短语in resonance作表语，译文中省略介词in。例13中的介词短语in the transmission of 在原文中作状语，译文中省略介词in。

(4) 英译汉时可省译动词。

英语句子的谓语必须用动词，汉语不仅可以用动词作谓语，还可以直接用名词、形容词、偏正词组、主谓词组等作谓语。因此，汉译时往往可以省略原文的谓语动词，使译文通顺、简练。

例14 Evidently semiconductors *have* a lesser conducting capacity than metals.

显然，半导体的导电能力比金属差。

例15 This laser beam *covers* a very narrow range of frequencies.

这个激光束的频率范围很窄。

例16 With regard to their products, they *have* the same concerns of the interrelationship of design, materials, and manufacturing process.

关于产品，他们同样要考虑其设计、材料和制造工艺之间的相互关系。

例14中的动词have省译，译文中采用"的"来连接偏正结构。例15中的动词covers省译，译文中采用形容词"很窄"作谓语。例16中的动词have省译，原文的宾语concerns转译为谓语"考虑"。

(5) 英译汉时可省译冠词。

冠词是英语名词前常使用的虚词之一，在其本身没有实际意义时，可省译。

例17 The resistance of *a* conductor is closely related with its length, its cross sectional area, and *the* material of which it is made.

导体的电阻与其长度、截面积和制造材料密切相关。

例18 *The* basic function of *the* triode is as an amplifier.

三极管的基本功能是用作电流放大器。

例19 *The* greater *the* resistance of *a* wire, *the* less electric current will pass through it under *the* same pressure.

在电压相同的情况下，导线的电阻越大，流过的电流就越小。

例17—19中分别省译了斜体冠词a与the。

(6) 英译汉时可省译引导词there, it。

首先，在英语中，以there作引导词的句子使用范围相当广泛，可以与be动词以及其他不及物动词，如seem, appear, exist, happen, stand, remain等连用，构成与汉语不同的一种"某地存在/有"句式。在这种句式中 there 已经失去了原有的意义,可省译。

例20 *There* exist neither perfect insulators nor perfect conductors.

既没有理想的绝缘体,也没有理想的导体。

例21 *There* remains one more test to be carried out before putting the machine into operation.

在运行这台机器之前,还剩一项实验有待进行。

上述两例中 there 失去实际意义,译文中被省略。

其次,引导词 it 作先行主语或宾语,实际主语或宾语(动词不定式、动名词或从句)放在后面。这种结构中的 it 没有实际意思,可省译。

例22 *It* is a pity that the result of this test is not satisfactory.

遗憾的是,这次实验的结果不尽如人意。

例23 Scientists have proved *it* to be true that the heat get from coal and oil originally from the sun.

科学家已证实,我们从煤和石油中得到的热都来源于太阳。

例24 *It* was not until the middle of the 19th century that the blast furnace came into use.

直到19世纪中叶,高炉才开始使用。

以上三例中 it 均被省译。

(7) 英语中有些同义词或近义词往往可以连用,或者表示强调,使意思更加明确,或者表示一个名称的不同说法。在英译汉时,往往省略其中一个词。

例25 The mechanical energy can be changed back into electrical energy by means of *a generator or dynamo*.

利用发电机可以把机械能转变成电能。

例26 Insulators in reality conduct electricity, *but, nevertheless,* their resistance is very high.

绝缘体实际上也导电,但其电阻很高。

例27 It is essential that the *mechanic or technician* understand well the characteristics of battery circuits and cells.

技术人员很好的了解电池的电路特性及连接电池的适当方法是至关重要的。

例25中的 generator 与 dynamo 同义,省略其中一个,译为"发电机"。例26中的 but 与 nevertheless 同义,译为"但"。例27中,mechanic 与 technician 在句中意思相近,译为"技术人员"。

(8) 根据汉语习惯,译文中可以省略一些意义已经隐含在句中的词。

例28 Even though assaults on the planet's ecosystem started a long time ago, *the period* of greatest destruction to the earth began after 1945.

即使对地球生态系统的冲击是在很久以前就开始了,但最严重的破坏是还应该说是在1945年以后。

例28中斜体部分所表达的时期已经隐含在"after 1945",因此,汉译时省译。

从以上各例中不难看出,删减某些词语才会使译文简洁、流畅、自然,而且突出了原文的整体意思。从原则上讲,翻译不能对原文内容作任何删减,所以,采用省略法必须以不损害原文的内容为前提。

五、分译法

英汉两种语言的句子结构区别很大。英语重形合,句子结构主要靠语法手段进行搭建,结构严谨。句子以主语、谓语为基础,借助介词、连词、连接词或短语、关系代词和关系副词等层层严密地把句子串起来,句中只能有一个谓语动词。汉语重意合,句子结构往往按照句内的逻辑、时间、空间和心理等顺序层层展开,一个句子可有多个动词连用。因此翻译英语时,需要把原文的某些成分单独拿出来翻译,译成汉语独立的句子,使译文意思简明、层次清楚,符合汉语表达习惯,这就是分译法。

1. 英译汉单词分译

(1) 分译副词。

例 1 Stylus is *essentially* a voltmeter.

实质上,触笔是个电压表。

副词 essentially 被前置出来,与主干分离。

(2) 分译形容词。

例 2 There are some partly mechanized versions of the process, however, in which *higher* current and deposition rates can be used.

但是这种(焊接)方法有些已经部分地机械化了,所以能使用较强的电流和较高的焊着率。

例 3 He spoke with *understandable* pride of the invention of the instrument.

他谈到那种仪器的发明时很自豪,这一点是可以理解的。

原文的形容词 higher 分译成"较强的"和"较高的"修饰电流和焊着率。例 3 中形容词 understandable 分译成了汉语复合句中的子句。

(3) 分译名词。

例 4 Nutche filter has the advantage of *simplicity* of construction and operation.

Nutche 过滤器的优点是结构简易和操作方便。

例 5 When you turn a switch, you can easily operate lighting, heating or power-driven electrical *devices*.

旋动开关,就能毫不费力地启动照明装置,供热装置和电动装置。

例 4 原文中的 simplicity 分别译成"简易"和"方便"。例 5 中,devices 分译成三个"装置",此为重复分译。

(4) 分译动词。

例 6 For example, elasticity, resistance to rupture and abrasion, durability, rigidity, and particularly mechanical stability are *improved* as a result of vulcanization.

例如(橡胶)通过硫化处理,能提高弹性,增强抗断裂性与耐磨性,延长使用寿命,增加硬度,尤其是可以改善机械稳定性。

例 7 Tests for gas filters are properly carried out in the design stage and also on the complete filter. They are described in various *documents published* by bodies as the U.S. Atomic Energy Commission, the United Kingdom Atomic Energy Authority, Undertakers'

Laboratory, and by manufacturers.

对于气体过滤器,应在设计阶段就进行适当试验,还应对整个过滤系统进行试验。试验方法在某些机构(例如美国原子能委员会、联合王国原子能管理局、保险商试验室)所公布的文件中和一些制造商所发表的资料中均有描述。

例 6 中的动词 improved 分别译成"提高""增强""延长""增加"和"改善",用以搭配不同的宾语,从而使语句通顺,符合汉语表达习惯。例 7 中的"文件"和"资料"是名词 documents 的分译,"所公布的"和"所发表的"则是动词分词 published 的分译。

2. 英译汉短语及分句的分译

(1) 分译介词短语。

例 8 The resistance of any length of a conducting wire is easily measured *by finding the potential difference in volts between its ends when a known current is flowing*.

已知导线中的电流,只要求出导线两端电位差的伏特数,就不难测出任何长度的导线的电阻。

例 8 中的介词短语 by finding the potential difference in volts between its ends when a known current is flowing 被分译成句子"只要求出导线两端电位差的伏特数"。

(2) 分译分词短语。

分词短语往往有附加的状语含义,因此,翻译时通过分析可将时间、条件、原因、目的、结果等从属意义表达出来。

例 9 *Taking the example of* a microphone as initial source again typical source impedance will be around 100 Ohms.

仍以麦克风为初始信号源为例,典型的源阻抗在 100 欧左右。

原文中的分词短语 taking the example of 译成表示条件意义的分句"仍以麦克风为初始信号源为例"。

例 10 The signal should be filtered before *being amplified*.

放大信号前,应先对其进行滤波。

原文中的分词短语 being amplified 译成表示时间的分句。

例 11 A diode placed in a circuit acts like a valve, *allowing current to flow in only one direction*.

装在线路中的二级管象阀门一样,只允许电流按一个方向流动。

原文中的分词短语可分译成句子,译为"允许电流按一个方向流动"。

(3) 分译动词不定式。

例 12 Many practical amplifiers chain together a series of analog amplifier stages *to obtain a high overall voltage gain*.

许多实际的放大器将多个放大器级联起来,以获得较高的电压增益。

原文中的动词不定式 to obtain a high overall voltage gain 分译成结果偏正复句中的结果分句。

例 13 The capacity of individual generators is larger and larger *to satisfy the increasing demand of electric power*.

单台发电机的容量越来越大,目的就是满足不断增长的用电需求。

动词不定式 to satisfy the increasing demand 分译成目的偏正复句中的分句,表示目的。

(4) 分译状语成分。

例 14 Science and technology modernized, industry and agriculture will develop rapidly.

如果科学技术现代化了,工农业就会迅速发展。

根据逻辑关系,原文中的状语 science and technology modernized 分译成汉语假设关系复句的假设从句。

(5) 分译名词化结构。

例 15 The signal levels inside power amplifiers are so much larger than these weak inputs that *even the slightest "leakage" from the output back to the input may cause problems*.

功率放大器中的信号幅度比微弱的输入信号大得多,即使输出的极微小的泄漏传输到输入端,都会引发一些问题。

根据逻辑关系把原文中的名词短语 even the slightest "leakage" from the output back to the input 译成让步复句中的从句"即使输出的极微小的泄漏传输到输入端"。

例 16 The increase in water pressure with depth makes it difficult for a man to go very far below the surface.

由于水压随深度增加而增加,人们要到水下极深处去有困难。

按原文的内在联系,将上句中谓语前的名词短语译成表示原因意义的分句,使译文易读易懂。

例 17 *The wrong power-line connections* will damage the motor.

如果把电源接错,就会损坏电动机。

从原文的句义看,名词短语 the wrong power-line connections 表示假设关系,因此分译为结果复句的分句"如果把电源接错"。

(6) 分译定语从句。

例 18 In power stations, coal or oil is burned and changed into electricity, *which is then sent over to homes and factories*.

在发电站里,煤和油燃烧变成电,然后电沿着导线输送到住家和工厂。

原文的定语从句分译成汉语的单句"然后电沿着导线输送到住家和工厂"。

3. 英译汉长句的分译

汉语多用短句,翻译时可把长句中的从句或短语转换为句子,分开来翻译。

例 19 The diode consists of a tungsten filament, *which gives off electrons when it is heated, and a plate toward which the electrons migrate when the field is in the right direction*.

二极管由一根钨丝和一个极板组成,钨丝受热时就放出电子;当电场方向为正时,这些电子就移向极板。

例 20 *Applied science, on the other hand, is directly concerned with the application of the working laws of pure science to the practical affairs of life, and to increasing* man's control over his environment, thus leading to the development of new techniques, processes and machines.

应用科学则直接涉及应用问题。它研究的是如何将纯科学的工作定律应用于实际生活,应用于加强人类对其周围环境的控制,从而导致新技术、新工艺和新机器的产生。

通过上述分析可见分译是有限定范围的,并不是要把句子弄得七零八落,支离破碎。分译后的汉语应该主谓结构完整,层次分明,逻辑概念清晰。否则分译的意义也就不复存在了。

六、合译法

合译法是在确保如实转达原文含义的前提下,将原文的并列句、复合句等复杂句型压缩成简单句或将语篇中两个独立句子合并成一个句子,以使译文更加紧凑、通俗易懂。合译的操作程序主要是合并、删除、移位、完形。合并是必须的、强制性的程序,删除是可选而且是频繁使用的技巧,移位的频率略低,完形是偶尔用到的技巧。在语义单元之间衔接时,有时候需要增词,以满足形式衔接的需要。

(1) 英语单句的合译。

单句是由短语或单个的词构成的句子,即没有分句的句子。单句的合译,指把一种语言的两个或两个以上相邻的且有句号隔开的简单句合并译成另一种语言的一个单句,或包含两个或两个以上分句的一个句子。英语单句的结构一般分两种:主语+谓语+宾语;主语+系表结构。英语单句的合译是将不同的英语单句成分组合在一起,使其更符合汉语言简意赅的表达方法。合译的两个单句相邻,而且需在意思上紧密相连,像补充说明、前因后果、并列存在或条件递进等。

例1 The coal-fired generating units have achieved an overall availability of 94% this year, and the forced outage rate has remained extremely low at 1.07%. This corresponds to the highest world standards and shows up the high quality of the power plant and its operation.

在本年内燃煤发电机组的整体可用率达94%,强迫停机率极低,仅1.07%,这与世界最高的标准相吻合,显示出机组具有很高的运行水平。

原文中的两个单句是典型的前后紧密联系句式,后句对前句进行补充说明,句意完整。前句表达的是燃煤发电机组运行的各项具体数据,后句补充说明这些数据的意义在于体现了发电机组的较高运行水平。因此,我们把英语中两个联系紧密的句子合译为一句,简洁明了,意义完整。

例2 This plant is located in the southeast of Jilin Province. It is one of the largest power plant in the province.

这个电厂位于吉林省东南部,是该省最大的电厂。

原文中的两个单句有共同的主语,并指代的是同一事物,但内容上并不相关。前句表明电厂的厂址,后句表明该电厂是最大的电厂,这两句为并列单句。因此,在翻译时为了避免形式上的主谓重复,需要合译为一句。

(2) 英语主从复合句的合译。

英语中的复合句,在汉语中又叫复句,指的是由连接词连接的一个或一个以上句子的长句。所谓复合句的合译,指把原文由两套主谓结构组成的复合句合译成汉语的只有一套主谓结构的句子,即简单句。这一译法主要是为了适应汉语的表达特点,使译文简洁、重点突出。

主从复合句由一个主句和一个或一个以上的从句构成。主句为句子的主体,从句只用作句子的一个次要成分,不能独立成为一个句子。从句通常由关联词引导,并由关联词将从句和主句联系在一起。从句按其在复合句中的作用分为主语从句、表语从句、宾语从句、定语从句和状语从句等。在主从复合句的合译中,先要分析出句子属于哪一类的主从复合句,再根据各类从句不同的特点进行合译。因此,英语主从复合句在合译时,应做适当删减或增添,把英语长句子压缩为汉语简单易懂的短句子。

例3 *It is apparent* that the design engineer is a vital factor in the manufacturing process.

设计工程师显然在制造过程中起着主要的作用。

在例句中,主从复合句合译为简单句。其中,主语从句部分翻译为主句,而 it is apparent 则译为"显然",在句中作状语,与从句重新组合为简单句。

例4 The circulating water system of this power plant is a closed-cycle type of cooling system *which* consists of cooling tower, basin, pump house and supply/return pipelines.

该电厂循环水系统是由冷却塔、水池、泵房以及供水/回水管路组成的闭式循环冷却系统。

原文是由 which 引导的定语从句,在合译时,删除了不必要的词,并调整了句子的顺序,更便于汉语的理解。

(3) 英语并列复合句的合译。

两个或两个以上的简单句用并列连词连在一起构成的句子,叫做并列句,其基本结构是"简单句+并列连词+简单句"。并列连词有 and, but, or, so 等。并列句中的各简单句意义同等重要,相互之间没有从属关系,是平行并列的关系。它们之间用连词连结。不同的并列连词表示并列分句之间的不同关系。

英语并列复合句是由两个或多个简单句和一个或多个从句构成的句子。这种句子容量大,可以提供更大的自由表达度。并列关系的句子连接的方式为分号连接或逗号加连词(and, or, but, for 等,用以表达两个分句间的逻辑关系)。被连接的两个句子可以是简单句,也可以是主从复合句,甚至是并列复合句。构成并列句的两个分句应该有逻辑联系,不能毫不相干。

例5 Use a transformer *and* power at low voltage can be changed into power at high voltage.

低电压通过变压器转变为高电压。

例5是由 and 连接一个祈使句和简单句合成的,在翻译时,用"通过变压器"替换祈使句,把原文合译为一个简单句,简单而易懂。

例6 The amount of power being used changes during the day, *and* the job of the control engineer is to switch power stations in and out.

一天内用电量是变化的,因而调度人员的工作就是让一些电厂投入或退出运行。

例6是由两个主语不同的简单句 and 连接而成的,在翻译时,分清两句之间的关系,合译为因果关系的两个分句,这样逻辑更加清晰。

例7 Insulators in reality conduct electricity *but* their resistance is very high.

绝缘体实际上是大电阻的导电体。

原文是由 but 连接的两个转折性意义的并列复合句。虽然两句话主语各不相同,在翻译时也可以按照逻辑关系,适当省略添加,组合成一个简单句。

以上我们所介绍的各种翻译技巧须灵活应用,无论是采用哪一种方法,其目的都是为了准确流畅地用译文表达原文的思想内容,从而服务读者。事实上,在翻译实践中,我们将根据英语文本的文体特征,结合翻译理论与翻译原则,选择几种翻译技巧并将其有机结合,进行翻译活动,不会也不可能完全孤立地使用某一种翻译技巧。

第三章 能源电力工程管理的文本特征与翻译

翻译作为一个交流与转换过程,要求译者能够准确地表达译文所要表达的全部信息。这里所说的"信息"包括原文所含有的全部文字、思想、情感及形式等。翻译就像是在使用不同语言的人们之间的交流活动中搭建起了一座桥梁。但要做到这一点并非易事,不是只要懂外语就能做翻译。要想真正地做好翻译工作,除了懂外语,还必须掌握一些翻译标准和翻译技巧,不断提高语言文字水平以及文化、专业知识水平,并在实践中不断地磨练。本章详细介绍能源电力工程管理的文本特征与翻译标准等。

第一节 能源电力工程管理文本特征

能源电力工程管理涉及范围广泛而复杂,其专业分工细,同时还涉及诸多相关领域。能源电力工程管理文本所述的专业知识深奥,选用的词汇生僻,语言表达结构复杂。因此了解其特征对能源电力工程管理文本的翻译具有重要意义。

无论能源电力工程管理英语文本还是能源电力工程管理汉语文本,都属于科技文体。随着科学技术飞速发展,科技文献大量出现,逐步形成了一个风格独特的文体,这引起了语言学家的注意,开始对这个专用文体进行研究。

就语言而言,科技文体和普通文体同为一个语言系统。就词汇而言,科技文章中虽然有大量专业科技词汇,但普通词汇的比例要远大于专业词汇;此外,很多专业词汇和术语是由普通词汇转化而来的。就语法而言,科技文体同样没有自己独立的语法规则,使用的仍是普通语言的语法体系,只不过某种语法现象使用较多,带有科技文体的特色。

一、能源电力工程管理英语的词汇特点

1. 能源电力工程管理英语大量使用本专业词汇

1)普通词汇转化而成的专业词汇

能源电力工程管理英语专业词汇有许多是借用日常英语词汇,改变原词词义甚至词性将其转换为专业词汇。

例 1 Whichever way the two two-way switches are left, one of the wires is alive and the other is *dead*.

无论这两个双路开关合在哪一边,两根导线中总有一根是带电的,一根是不带电的。

原文中的 alive 和 dead 的基本义是"活着"和"死了",但在能源电力专业领域内就成了"带电"和"不带电"的意思。

2) 运用构词法形成的专业词汇

英语的构词法主要有合成、转化和派生三种方式。这三种方式在能源电力英语专业词汇的构成中都得到大量使用,其中派生法构成的词汇最多。派生词即通过加前缀、后缀构成新词。电力专业常见的前缀:表示"电"的 electr-, electri-, electro-;表示"磁"的 magnet-, magneto-;表示"水"的 hydro-;表示"热"的 thermo-;表示"自动"的 auto-;等等。后缀常常用来表示行为、性质、状态等抽象概念:-ment, -tion, -sion, -ance, -ence, -ure, -ity,等等。

2. 非专业词汇选词书面化

能源电力工程管理英语除了大量使用专业词汇,还包括了大量的普通词汇。这些非专业词汇大多偏向于书面化,大词和长词出现频繁。虽然普通英语或文学英语中也会使用非常正式的书面化的词汇,但它们在能源电力英语中出现的频率明显要高。例如,accomplish 代替日常英语中的 do;demonstrate 代替 show;magnify 代替 enlarge;numerous 代替 many;furthermore, therefore, however 等逻辑连接词代替日常英语中的 what's more, so, but 等。

能源电力工程管理英语采用正式书面语有利于提高文章的正式程度,体现其科技文体的特征。译者必须了解这一特征,在翻译时忠实于原文,译出原文的语言风格。

3. 能源电力英语广泛使用缩略语

缩略语表达简洁,拼写方便,广泛用于科技英语中,能源电力专业英语也不例外,行业内约定俗成的缩略语被大量使用。例如,alternating current(AC)交流电;direct current(DC)直流电;automatic generation control(AGC)自动发电控制;automatic voltage control(AVC)自动电压控制;wind power unit(WPU)风力发电机组。

4. 名词化

能源电力工程管理英语最重要的词法特征是名词化,主要体现为广泛使用由动词或形容词转化而来的抽象名词,或起名词作用的非限定动词(如 programming, processing 等)。名词化可以将一个句子缩减成一个词,使复合句简化成简单句,使语言结构简化,达到能源电力英语表达所追求的简练、浓缩的要求。另外,使用表示动作的名词有助于客观地表达所论述的概念。

例 2 Since the early 1950s, the *exploration* of nuclear fission technology on a large scale for the *generation* of electric power has resulted in many modern nuclear power plants.

自从20世纪50年代初期起,对用于发电的核裂变技术的广泛探索,使得许多现代化核电厂得以建立。

例 2 中的 exploration 和 generation 这些名词的使用使句子结构变得简洁紧凑,表现出良好的逻辑性与精确性,同时还增强了表达的客观性。

二、能源电力工程管理汉语的词汇特点

专业术语的大量使用是科技文体区别于其他文体的最主要特征之一。能源电力工程管理汉语是科技汉语的分支,大量使用能源电力专业词汇。例如,电流 current;电压 voltage;

电阻 resistance；电抗 reactance；电导 conductance；电纳 susceptance；等等。

作为科学术语，能源电力工程管理专业术语词义具有单义性，每个术语只有一个特定的精确的含义，否则会造成歧义，这是由科技文体严谨性特点决定的。同时术语又具有稳定性，词义一经固定下来，不会轻易变化。

三、能源电力工程管理英语的句法特点

1. 能源电力工程管理英语多用被动句

能源电力工程管理英语最突出的一个特征就是较多地使用被动句。这是因为能源电力工程管理人员所关注的是该领域里的客观事实和规律，所以能源电力工程管理英语文献的主要目的是客观而且准确地陈述事实、特征、报道科学发现、叙述研究方法等等。因此，能源电力工程管理科技文献通常使用非人称句作客观陈述，从而较多地使用被动句。

例3 Owing to the fact that electricity *can be transmitted* from where it *is generated* to where it *is needed* by means of power lines and transformers, large power stations *can be built* in remote places far from industrial centers or large cities, as *is cited* the case with hydroelectric power stations that are inseparable from water sources.

由于电力可以从发电的地方通过电线和变压器输送到需要用电的地方，因此大型电站可以建在远离工业中心或大城市的地方，离不开水源的水力发电站就常常是这样建立的。

原文中共用了五处被动语态：can be transmitted, is generated, is needed, can be built 和 is cited。

另外，英语中存在一些自动性很强的动词或动词词组，例如，originate, produce, enable, emerge, vary, give, send, occur, represent, flow, bring about, come 等，使用这些词或词组时，一般不用被动式。

2. 能源电力工程管理英语多用长句

由于要表达的科学事实或现象信息量大，电力专业论文常常需要使用很长的句子才能充分表达复杂的意思。长句在基本结构的基础上扩展，添加各种修饰语如定语、状语，使用并列句或并列成分如并列主语、并列谓语、并列宾语、并列定语或并列状语等，增加插入语、同位语或孤立语等，或者使用倒装和省略结构。往往并列结构、从句、非谓语结构交错使用，使句子变得错综复杂。在一个句子中有时会同时出现多个从向、多层并列结构等情况。

例4 Instruments that continuously monitor current, voltage, and other quantities must be able to identify the faulted equipment and to bring about operation of the circuit breakers which remove it from service, while leaving in service all other equipment on the operating system.

监控电流、电压和其他指标的仪表必须能识别有故障的设备，并使断路器开始工作，让有故障的设备停止运行，从而让运行系统中的其余设备仍继续工作。

原文有42个词，是一个标准的长句。主语 instruments 后带有一个 that 引导的定语从句；谓语部分结构复杂，to identify... 与 to bring about... 同为 be able 后的并列结构；名词 circuit breakers 后又有一个定语从句，句尾的 while leaving in service all other equipment on the operating system 是分词结构作伴随状语，其中 all other equipment on the operating system 是 leaving 的宾语，为了使句子平衡而将该宾语后移。

3. 能源电力工程管理英语多使用非谓语结构

由于能源电力工程管理英语要求简洁明了,追求言简意赅,所以非限定动词短语使用较多。因为各种非限定动词短语可以代替从句或分句,能大大缩短句子长度。

例 5 This system is economical as it flattens out the variations in the load on the power grid, permitting thermal power stations such as coal-fired plants and nuclear power plants that provide base-load electricity to continue operating at their most efficient capacity, while reducing the need to build special power plants which run only at peak demand times using more costly generation methods.

此系统很经济适用,因为它平衡了电力网上负荷的变化,使那些提供基本负荷电的热电站继续最高效地运行,如燃煤电厂和核电厂,也减少了修建特殊电厂的必要性,这些特殊电厂发电方法代价昂贵,只在用电高峰时运行。

原文为一个主系表结构,as 引导的原因状语从句中有两个分词结构作并列结果状语,分别是 permitting thermal power... capacity 和 reducing... methods。两个并列状语间用 while 连接,表示对比关系。

4. 能源电力工程管理英语中各种句型运用丰富

祈使句 It... to do 或 It... that...(主从句型)等句型在能源电力工程管理英语文本中较常使用。如果描述操作过程或试验过程等时,常用祈使句。

例 6 Note that power quality has nothing to do with deviations of the product of voltage and current(the power) from any ideal shape.

注意,电能质量和电压、电流乘积(功率)与任何理想形态的偏差无关。

例 7 Suppose that tube is large and smooth, the pressure makes many marbles flow.

假定管道巨大而光滑,其压力会使许多大理石子流动。

It... to do 或 It... that...(主语从句)句型在能源电力工程管理英语中使用也很多,用来表达看法、建议、结果、发现等。在这两个句型中 It 是形式主语,后面的 to do 或 that 从句是句子真正的主语。

例 8 It is necessary to make a great amount of effort to maintain an electric power supply within the requirements of the various types of customers served.

为了保证电力供应以满足不同类型用户的要求,必须付出很大的努力。

例 9 It is observed that when two current-carrying carbon electrodes were drawn apart and electric arc of intense brilliance was formed.

人们注意到,当将两个带电的碳极拉开时,就会产生极强的电弧光。

四、能源电力工程管理汉语的句法特点

1. 能源电力工程管理汉语多用主谓句

能源电力工程管理汉语是以描述电力领域科学事实、分析科学道理、描述科技发现等为目的的,所以行文要求逻辑严密、结构明晰、避免模糊和歧义。因此主语是客观现象或事实时,多用完全的主谓句。但如果主语泛指人时,则往往省略。

例10 The determination of the parasitic damping factor is a vital procedure in model solving and performance optimization for inertial coupling based vibration energy harvesting resonant micro generators.

寄生阻尼因数的确定对基于惯性耦合的谐振式振动能捕获微型发电机的模型求解、性能优化等至关重要。

2. 能源电力工程管理汉语少用被动句

能源电力工程管理汉语文本大多使用主动句,少用被动句。和英语大量使用被动语态来体现科技文体的客观性不同,汉语多使用主动语态阐释客观事实和规律。

例11 The control system is operated without involving any sequential decomposition process, and active/reactive powers of rectifiers output are stabilized. Steady-state and dynamic performance of the system is improved.

控制系统运行时无需对不平衡电流和电压进行正、负相序分解,并可同时抑制整流器输出的有功和无功功率波动,改善系统动态和稳态性能。

英文用了介词结构"对……进行分解"表示受动体,用主动形式表示了被动的意思。

3. 能源电力工程管理汉语单句结构复杂

能源电力工程管理领域的现象、规律错综复杂,并且随着能源电力工程管理科技的不断进步,人们对该领域的认识会不断深入细化,那么用来表达这些复杂关系的句子也同样会结构复杂。长而复杂的单句占很大比例。

能源电力工程管理科技文献中单句往往使用复杂的定语和状语成分,起到限定名词、动词、形容词的作用,使所要阐述的定语、定理、假说和现象等表述严谨、精确。如果不用这些附加成分,则往往造成概念上的模糊和对理论的误解。

例12 The feasibility of air-side heat transfer enhancement by staggered short wavy fins to flat tube in air-cooled condenser of power generating unit was studied by numerical simulations.

通过数值模拟的方法,对锯齿形短翅片应用于火电机组直接空冷凝汽器扁平管空气侧强化传热的可行性进行探讨。

中文为单句,但结构复杂。句中省略了主语"笔者/我们",谓语"进行"和宾语"探讨",构成句子的主干。使用介词结构"对……的可行性"做状语,而这个介词结构中的宾语"可行性"之前又有一个由一个句子构成的复杂定语,由主语(齿形短翅片)+谓语(应用)+补语(火电机组直接空冷凝汽器扁平管空气侧强化传热)构成。如果再深入分析,补语部分还是一个叠加的复杂定语。

五、能源电力工程管理英语的语篇特点

英语强调语法形式的统一和连贯,多用"形合法",通常使用衔接手段把句子和段落连接起来,形成语篇的连贯性。衔接手段是将句子连成篇章的纽带。根据英国语言学家韩礼德和哈桑的划分,英语衔接手段主要有五种:照应(reference)、替代(substitution)、省略(ellipsis)、连接(conjunction)及词汇衔接(lexical cohesion)(韩礼德,哈桑,1976)。能源电力英语主要使用其中的"词汇衔接""照应"和"连接"三种方式,另外两种则很少使用。

例13 ①An analysis of customer outages determined that about 78% of interruptions were due to faults on main three-phase lines. ②A fault on the main line would cause, on average, 2,000 customers to lose power. ③ In order to reduce these outages, the company has instituted reliability programs that include trimming trees, installing lightning arresters, installing covered wire and replacing obsolete armless insulators. ④These programs reduced outages, but were not sufficient to reach the desired level of service continuity. ⑤To further improve reliability, LILCO has installed an advanced distribution automation system, developed jointly by LILCO and Harris Distributed Automation Products, Calgary, Alberta, Canada. ⑥This system isolates faults and restores non-damaged portions of the main line circuit based on realtime parameters of voltage, current, breaker status and supervisor-controlled switch positions. ⑦Since this project was initiated in 1993, more than 240,000 customers have avoided a sustained interruption of service.

此段落是有关 LILCO 公司为解决断电的问题采用的解决方法。在这个段落中作者充分运用了词汇衔接这种衔接方式。第①句引出段落的主题：断电问题，出现主题词 outrage。这个段落原词复现出现多次，如文中反复使用了 outage 三次，program 两次，system 两次。interruption 和 lost power 是主题词 outage 的同义词复现。system 和 project 也是同义词复现关系。service continuity 和第④句中的 reliability 与 interruption, lose power, outage 互为反义，形成同现关系。此外，此段内容是关于供电线路断电问题的，与电路有关的词如 main line, circuit, wire, voltage, current, switch 形成词汇链，这些词在语义上的内在联系将句子连接成有机的整体，实现语篇的连贯。而且，这种词汇复现还起到了强调和突出主题的作用，从而加深了读者对篇章的理解。

例14 ①Hydropower has many advantages. ②First of all, hydropower generation is non-wasting self-replenishing and non-polluting. ③*It* is a physical phenomenon and no chemical change is involved. ④Water comes out unchanged from the turbine after imparting *its* energy and can be used again either for power generation or for irrigation. ⑤In fact, *this* is done in multipurpose river-valley schemes like the Chambal Valley development in India and the Tennessee Valley development in U.S.A. ⑥In the case of the Chambal Valley development, power is generated with the help of the *same* water in three powerhouses situated one after another on the river, before being released into irrigation canals. ⑦As against *this*, coal, oil or nuclear fuel can only be used once.

这个小段中作者多处使用照应这种衔接手段。第③句中的 it 回指前文中的 hydropower generation。第④句中的 its 指句首的 water，是句内照应。第⑤句的 this 概括前句内容，是指示照应。第⑥句的 same 是比较照应。第⑦句的 this 回指前文中的水力发电。

六、能源电力工程管理汉语的语篇特点

与英语不同，汉语强调"意合"，主要凭借词语、句子之间内在的逻辑联系形成语篇的连贯。因此，汉语缺少形式上的标志。能源电力工程管理汉语作为科技文体虽然比其他类型的文体使用更多的虚词表示逻辑关系，但"意合"这一特色在电力汉语篇章中仍有体现。在翻译成英语时需要酌情选用适当的连接词把这些语法关系明确表示出来。

例15 ①而且，对于什么是完美，仁者见仁，智者见智。②核能倡导者认为钚是一个非常理想的能源来源，因为许多民用和军用工厂都产生这种元素，只不过被军方堆积起来用来

制造武器,如果这些钚能在核反应堆内再燃烧而不是单单堆积存放,将会更好。③反对者认为,这种金属是现存的腐蚀性最大的物质之一,而且存放、处理、运输起来都非常危险,况且还可能被恐怖分子用来制造混乱。

在这个段落中,第②句和第③句在逻辑上是对比关系,表达了核能倡导者和反对者的不同观点,是对第一句"仁者见仁,智者见智"的具体说明。

第二节　能源电力工程管理翻译的标准

翻译标准作为翻译理论研究中最为重要的部分之一,关系到翻译理论研究、翻译实践和翻译事业的发展。翻译标准是进行翻译理论研究的根本基础,它可以反映人们对翻译的本质和翻译工作本身的认识。翻译标准不仅是制约翻译工作、衡量译作优劣的准则,也是评论译作的依据。

一、忠实准确

任何翻译作品都应忠实于原文,这是衡量翻译作品可信度的标准。这一翻译标准对任何文体的翻译都适用。所谓"翻译",就是用另一种语言如实地还原源语的本来面目。

翻译讲求忠实性、准确性、可信性,因而"忠实准确"对于能源电力工程管理翻译尤为重要。译文必须客观完整地表达、传递、重现原文的内容,原文和译文的指称意义一样,完全或者基本重合,即从概念对比的角度要求译文和原文表达的是同一概念,描述的是同一过程或者变化等。

能源电力工程管理技术资料的主要功能是论述科学事实、探讨科学问题、传授科学知识、记录科学实验、总结科学经验等,这就要求能源电力工程管理资料的标准首先必须忠实准确。所谓忠实准确,就是忠实地、不折不扣地传达原文的全部信息内容,要符合所涉及的科学技术或某个专业领域的专业语言表达规范(赵萱等,2006:2)。

要做到译文忠实准确,译者能够准确理解原文所表达的意思并且忠实地、不折不扣地传达原文的全部信息内容是第一步。

例1　The *power plant* is the heart of a ship.
动力装置是船舶的心脏。

power plant 是电力翻译中的常见短语,它的主要意思有两个,即发电厂和动力装置。要确定例1中 power plant 的确切含义,就需要译者根据上下文并结合一定的电力专业知识来进行判断。

二、通顺流畅

通顺流畅指的是译文文本流畅自如,译文在读者头脑中产生的印象和原文在读者头脑中的印象基本相同,译文读者能够产生跟原文读者基本相同的感受。这就要求译文的遣词造句和语篇构建简洁明了,重点突出、通顺易懂,符合译入语语法、句法和表达习惯,避免因为语言形式的差异而影响信息的理解。换句话说,在满足"忠实准确"标准的前提下,翻译出来的译文要地道流畅,能够为译入语读者所接受。这一标准也是顺应目的语语境选择的要求。违背这一要求,就会出现翻译腔。

例2 It should be realized that magnetic forces and electric forces are not the same.

译文1:磁力和电力的不一样是应该被认识到的。

译文2:应该认识到,磁力和电力是不同的。

译文1和2所传达的基本信息是相同的,但是译文1有"硬译"的现象,语言较为晦涩拗口,相比之下,译文2通顺易懂,更符合汉语的习惯。

例3 It is forbidden to dismantle it without permission so as to avoid any damage to its parts.

译文1:为了避免损坏设备的零件,未经许可不得拆卸该设备。

译文2:严禁乱拆,以免损坏该设备的零件。

译文1和2比较起来,译文2简短精炼、一目了然,能够更好地起到警示作用。

三、规范专业

规范专业要求译文的专业术语表述符合科技语言和术语的规范,尽可能利用译入语中已有的约定俗成的定义、术语和概念。本书讨论的译文读者是指有相当认知能力和教育文化水平,并熟知其所在行业的相关专业术语和惯用语的理想化读者。因而,要使这样的读者接受和认同译文,译文就必须要达到规范专业的要求,这就要求译者具备一定的科学技术知识,或者能够通过查询专业资料和求教于相关领域的专业人士,透彻理解原文内容,根据文本正式程度的要求,尽可能译出规范、专业、约定俗成的定义、概念和术语。当今科学技术发展很快,翻译人员需要注意及时给自己"充电",从而适应当下的科技发展形势,使自己的译文符合专业领域的语言表达规范。

能源电力工程管理翻译的译文也要符合行业专业语言表达规范。

例4 The various types of *single phase a.c. motors* and *universal motors* are used very little in industrial applications, since *polyphase a.c. or d.c. power* is generally available.

因为通常有多相交流电源或直流电源可用,所以在工业上很少使用各种类型的单相交流电动机和交直流两用电动机。

例5 在不需要电绝缘的地方,有时使用自耦变压器。

Where there is no need for *electrical isolation* an *auto-transformer* is sometimes employed.

例4和例5中涉及single-phase a.c. motors、universal motors、polyphase a.c. or d.c. power、电绝缘、自耦变压器几个能源电力的专业术语,译者要通过查阅专业词典或专业书籍等严谨的方式把它们翻译成行业通用的规范化术语——单相交流电动机、交直流两用电动机、多相交流电源或直流电源、electrical isolation 和 auto-transformer。

四、逻辑清晰

能源电力工程管理文献反映的是逻辑思维的结果,其特点是逻辑严密、概念清晰、表述无懈可击。因此,译者在翻译过程中必须充分考虑句中的各种语法关系和概念之间的逻辑关系,结合上下文对句子结构进行综合分析,最终用符合逻辑的语言传达出原作者的思想和意图。

例 6 *Where the volt is too large a unit*, we use the milli-volt or microvolt.

如果伏特这个单位过大，我们就应该用毫伏或微伏。

原文中 where 引导的从句从英语语法的角度分析是一个典型的地点状语从句，但在逻辑意义上相当于一个条件状语从句，因此在译成汉语时按照条件从句来译，则在逻辑上显得更加通顺。

例 7 为了可以将储存于燃料中的化学能量转变成有用功，首先需要通过燃烧把化学能转变为热能。

It is necessary first to convert the chemical energy into heat by combustion *for the purpose that useful work from the chemical energy stored in fuels might be produced*.

汉语习惯于将目的状语放在句首，而译成英文后将 for the purpose that... 放在了句子的后半部分，这样在逻辑上更符合英语的表达习惯。

五、服务读者

翻译目的论学者认为翻译行为所要达到的目的决定整个翻译行为的过程，因此译者首先应该明确翻译行为在特定的翻译语境中的特定目的。在前文提到的目的论概念的框架下，受众就成了决定翻译目的的重要因素。受众——译文预期的接受者，他们有自己的文化背景知识、对译文的期待以及交际需要。目的论确立了译者的主体作用，强调译者在翻译过程中要以译文的预期目的和功能为出发点，了解受众的特点及需求，进而灵活的选择相应的翻译策略。

目的论认为翻译是在目的语境中为某种目的及目的的受众而产生的语篇，每一种翻译都指向一定的受众。专用能源电力工程文本包括基础理论论著、专利文件、技术标准、技术合同、科技论文等。其特点是表达客观、逻辑严密、行文规范、用词正式、句式严谨。普通能源电力工程文本包括操作规程、维修手册、安全条例、产品说明书、产品促销材料、科普读物等。其特点是用词平易、句式简单、多用修辞格。这两种类型的文本的受众也是有区别的，其中专用文本主要是科学家之间、高级管理人员之间、同一领域的专家之间使用；普通文本主要是生产部门的技术人员或工人之间、生产商与消费者之间、专家与外行之间使用。

由此可见，译者在翻译能源电力工程文本的时候要明确译文的文本类型，进而确定受众类型，然后综合这些要素发挥自己的主观能动性，选择最佳处理方法，最终呈现针对性强的翻译作品。

第三节 对译者的要求

翻译是由译者主导的在不同语言之间进行的一种语言交流活动。如前文所提到的，译者所起到的作用就好比在使用不同语言的人们之间架起了一座桥梁，把一种语言文字所表达的意义用另一种语言文字表达出来。这种语言转换的过程要求译者能够准确地表达译文所要表达的全部信息。要想真正地做好翻译工作，除了懂外语，还必须掌握一些翻译理论和翻译方法，不断提高语言文字水平以及文化、专业知识水平，并在实践中不断地磨练（赵萱等,2006:1)。要做好能源电力工程管理翻译工作,译者应该达到以下几项基本要求。

一、基本素质

1. 语言能力

我国21世纪翻译人才的任务不仅是把国外的先进东西翻译进来,而且还要把我国的优秀文化、科技成果推向世界,让世界了解我国的民族文化和科技发展。对于从事翻译工作的职业人员来说,具备扎实的语言功底是翻译人员必备的基本素养,无论是汉语功底还是外语功底,都需要经过长期的专业学习和不断的经验积累方可取得,可以说,语言素质是衡量职业译员的基本标准之一。

翻译涉及两种语言的转换,对译者语言素质的要求自然很高。这里的语言素质首先包括对源语的理解和对目的语的运用能力。译员的双语能力不仅仅是指通晓基本语言知识(语音语调、语法结构、词法语义等),更重要的是指运用语言知识的能力(即听、说、读、写、译的能力),但是并非懂得双语的人就能成为一名合格的译员,换言之,掌握两种语言只是成为一名合格译员的必要条件,而不是充分条件。一般来说,驾驭双语能力既是译者语言素养的体现,也是衡量译者翻译水平的一个重要标尺。此外,语言与文化是不可分离的。一个合格的译员,不仅要有驾驭双语语言的基本能力,还必须了解两种文化。文化的差异无论什么时候都是存在的,而翻译人员的职责,就是要缩小这些差距,如果译员对两国文化不甚了解,就很难准确地翻译出地道的译文。所以说,要成为一个合格的译员,首先要掌握或精通两种语言文化。

译者的语言素养在能源电力工程管理翻译中同样具有重要的作用,换言之,过硬的语言能力是进行能源电力工程管理翻译的首要条件。如果说翻译既是科学又是艺术,那么不可否认,能源电力工程管理翻译的科学性高于艺术性,但要准确地传达这种科学性,过硬的语言能力是不可或缺的。这首先表现为在翻译过程中译者对外语和母语两种语言知识的掌握程度。译者一方面要正确理解原文,将原文本来的内容严谨、准确地表达出来,保证译文的忠实性;另一方面要用通顺、合理,且符合目标语习惯和科技语言规范的术语表达出来。众所周知,科技文体具有其特殊的功能性。所谓"失之毫厘、谬以千里",如果译文晦涩难懂、含糊不清,让读者不知所云甚至误解,那么译文的功能性就大打折扣了。

能源电力工程管理翻译中的语言素质不仅要求译者对双语语言知识的掌握,还要求译者在翻译过程中具有一定的语言敏感性。严谨的翻译态度能够帮助译者查错纠错,而过硬的语言能力同样具有相似的功能。语言能力不足的译者易被原文复杂的文字和句式所束缚和迷惑,翻出来的译文拗口不顺,可读性差,给读者带来阅读困难,甚至译文中出现违反科学常识或专业常识的地方自己也浑然不觉。相反,语言能力较强的译者具有一定的语言敏感性,能够辨识不通顺、不合逻辑的译文,并从中找出译文中可能存在的问题。此时,即使译者由于受科技知识的限制,尚不能确定正确的译法,语言的敏感性亦能让其发现问题、提出问题,然后通过查阅资料、学习相关知识解决问题,从而避免、纠正误译。从某种程度上说,过硬的语言能力能在翻译过程中帮助译者弥补科技知识上的弱势,而这是很多人所忽略的。

能源电力工程管理翻译的对象具有学科专业性和知识前沿性的特点。这种特殊性对译者提出了更高的要求和挑战,要求译者既要通晓两种不同语言,又要具备一定的学科相关知识。准确性、客观性、逻辑性、严密性、连贯性、简洁性是科技类文献的基本特征。要将这些特性在目标语言中完整再现出来,译者首先要在语言层面上准确分析原文的结构含义、上下文逻辑关系;同时,科技文体在功能上起到传达原文作者科学思想和传播科学知识的作用,这就要求译者用通顺、地道、专业的科技语言将译文准确表达出来,即首先要做到忠实与通顺,而要做到这一点,译者的语言素质是不容小觑的,相反,如果连忠实和通顺都做不到,翻译便失去了它的目的和意义。

译者的语言素质主要包括两个部分:本族语的语言素质和外国语的语言素质。本族语是做好翻译工作的前提和基础,人们大都有一种误解,认为只要学好外语,从事翻译工作就没什么问题,但事实并非如此。提高本族语水平,对翻译而言就如同盖房子打地基一样,基础越坚实房子才能盖得越高,本族语的基础打得越扎实,外语水平才能提得越高。对于中国译者而言,汉语是母语,理应没有问题,但事实并非如此,有些翻译工作者,能够说一口比较流利的英语,但是一说汉语反而词不达意、不合逻辑。所以,翻译工作者必须要经常有意识地加强母语学习,夯实基础,提高对母语的理解、运用和表达能力。外宣工作中,译员更要善于仔细、深入、准确地理解中文内涵,唯有如此,才能保证译文的准确和质量。

另外,翻译人员的外语能力至关重要,这一点不难理解。以英语为例,译者的能力至少应体现在如下几个方面:首先,全面地掌握英语语法知识。语法是一门语言的规律总结,掌握了英语语法,就意味着对该语言有了总体的认识,也就是获得了一把学好英语的钥匙。语法对于规范译文具有指导意义,尤其是专业性较强的科技英语(包括能源电力工程管理英语),这是因为科技材料大多是结构严谨、用词正式的书面语。其次,要具有丰富的英语词汇量。丰富的词汇量会大大提高翻译的速度和准确性。英语的词汇往往一词多义,有些词在词典里虽有某个汉语所表达的意义,但实际上并非是最佳的表达方法。而词汇量丰富的译员能根据目的语的语法和修辞特点,能够用地道的目的语词汇准确地表达出原文所要表达的含义。最后,英语能力还体现在对英语文化知识的掌握上。译者不仅要掌握基本的英语语言知识,还应尽可能多地掌握其他一些与英语语言有关的知识,如英语修辞、文体、语用、历史、文化、文艺等,这方面的素养也会对翻译工作具有重要的指导意义。在翻译实践中,由于译者的语言功底欠佳而造成的错译、误译屡见不鲜,比如,将"把中国建设成为"(turn China into)译成"build China into"等等。可见,翻译工作人员的语言水平对于提高译文质量具有至关重要的作用。

既然语言基本功是翻译人员应该具备的基本素质,那么怎样培养和提高翻译人员的语言素质呢?

第一,提高语言对比和分析能力。

从语言学的角度来看,翻译是不同语言间的符号转换和信息传递,而转换是在语言对比中进行的。乔姆斯基认为,不同语言的基本规则在很大程度上是具有普遍性的,尽管不同语言的表层结构各不相同,但是深层结构都是相同的。换言之,不同的语句表象都是因为相同的深层结构经过不同的转换而变得形式各异。在翻译过程中要克服源语和目的语结构上的差异,比较两种语言各自不同的转换规则和表现形式。从思维反映现实的观点来看,事物的主体(语法上通常作主语)发出动作(通常为谓语),这个动作又涉及另一事物(作宾语)。事物总是在一定的时间、地点、条件下运动和发展的,因此在语言表达上需要状语,事物本身又有不同的性质与特征,这就需要定语等等。翻译不免要在词、词组、分句、句子、段落、篇章等各个语言层面上进行比较。例如,从复句表达的次序来看,汉语习惯于按事情本来的程序,先偏后正，先因后果，先假设后论证，先让步后推论等等，即先从后主，而印欧语往往先主后从。再如在词组的层面上，汉语常用四字结构，表现形式与印欧语有很大差异。通过深入对比可以发现，源语和目的语在思维方式、语言结构、修辞手段等方面同中有异、异中有同。总之，对两种语言掌握得越熟练、越细致，比较得越深入、越全面，在表达上就越精确、越流畅，译文就越准确、越地道。语言分析是翻译工作（特别是笔译）的一个重要步骤。英语以长句、复合句见多，结构复杂、逻辑性强。碰到要解释的名词，随处都可以加以解释，这就造成作定语的短语、从句偏多，增加了句子的复杂性；碰到要修饰的词，又随处可以加以说明，这就产生了作状语的短语、从句等等。但是无论句子怎样变化，多么复杂，句子成分总不外乎主语、谓语、宾语、补语、状语、定语之类。句法结构的分析方法也大致相同，先确立主语，再理出分支，顺藤摸瓜，因势利导；分析句子时需化整为零，翻译时再化零为整。当然分析时切不可处处以汉语比附，以致牵强附会，不得要领。如汉语中被说明语和说明语是前后紧挨着的，英语的定语有前置和后置之分，状语则可置于句首、句中、句末。英语被动式用得较多，当不需要说出行为主体时，或强调行为所及的客体时，当句子结构更便于安排时，均可用被动式，但在汉语中一般只是在强调宾语时才用，因此汉语句式多主动，英语句式多被动。总之，分析语言的目的是为了准确地理解语言，同时也是为了更好地表达语言，在实际翻译过程中，离不开对两种语言的对比和分析。

第二，提高语言的表达能力。

具有较强的语言表达能力是从事翻译的基本要求。要把原文的意思确切地表达出来，要求译者选用适当的语言材料和采用合乎目的语规范的句法手段把它组织起来。口头翻译要求迅即反应，脱口而出，笔头翻译虽有充分的时间推敲，但是如果缺乏语言修养，冥思苦想也难得佳句，逐字死译或者杜撰词语又会损害语言或破坏原意。而且，不同文体有不同文体的表达方式。

母语是做好翻译工作的基础。扎实的母语基本功有利于外语水平的提高。母语水平主要包含两个方面的能力：母语的理解能力和母语的表达能力。汉译英时，尤其需要很强的母语理解力。反过来，在英译汉时，如果只有较强的英语能力而没有较强的母语表达能力，那就很难将源语中一些精彩的内容用同样精彩的母语表现出来，翻译的目的便无法实现。只有对母语具有较为全面、深入的了解，具有很强的理解能力和表达能力，具有深厚的语文知识，才能在翻译理论研究和翻译实践中发现和找出两种语言共有的规律性和两种语言在词汇、语义、结构、思维方式、逻辑推理等方面的差异性。遇到问题时，才能知道从何处着手去解决，从而达到顺利完成翻译任务的目的。翻译工作对外语的要求也同样包括语言理解能力和语言表达能力。作为一名合格的译员，必须掌握大量的词汇、习语、谚语、俚语，能够灵活熟练地运用语法手段和修辞技巧。只有这样，才能用不同的词汇表达同一概念，用不同的方式表达同一思想内容，然后从中选择最合适的词句和最恰当的表达方式。这两方面的能力加强了，才能在外译汉时正确地理解原文，掌握原文的主题思想和写作风格，在汉译外时，才能用道地的外语忠实、流畅地进行表达。

2. 专业知识

在我国翻译行业中，从业人数最多的是科技类文本的翻译工作者，与目前市场经济联系更直接、最紧密的也是科技类文本的翻译工作者。然而，部分科技类文本翻译的质量却令人担忧，造成这种现象的原因主要有二：一是翻译人员对所译领域的专业知识知之甚少，二是相关领域专业人才的外语翻译水平相对薄弱。简而言之,既能翻译又懂专业的人才非常匮乏。而专业知识在翻译中是非常重要的,如果说文字功底是文学翻译的基石,那么专业知识则是科技类文本翻译的基石。俗话说隔行如隔山,如果译者对所译资料的专业知识一无所知或知之甚少,那么他或她就不可能得心应手地进行翻译,其译文极有可能会出现不准确、不充分或不地道等问题。

译者的专业知识主要是指译者应具备的翻译基础理论知识和翻译技巧以及翻译所涉及的相关领域的专业知识。所谓相关领域的专业知识,主要是指人们在某专业领域内所掌握的知识,而且往往需要经过长期的学习和培训才能掌握。不少人认为,翻译工作者没有专业。这句话应该辩证地看待,如果说它不正确,是因为从事翻译工作的人员也需要经过翻译专业的学习和培训,这本身就是一种专业;如果说它正确,是因为译者往往根据需要翻译不同专业的材料,或政治,或贸易,或科技,或文学,没有固定的专业。然而,对于译者而言,专业知识旨在帮助正确理解和表达原文,使译文更加准确,因此,翻译人员所要求的专业知识不需要像专业人员那样精深,只需要掌握所译专业的基础知识、基本原理和基本术语即可。事实上,翻译人员在工作中遇到的翻译难题,并不是因为缺乏外语知识,而是因为对原文所涉及的专业知识缺少必要的了解而造成的。另外,由于一个人的能力和精力毕竟有限,不可能同时掌握多门专业,只能根据自身或翻译的需要学习或了解一两门专业,以保证翻译的质量。

能源电力工程管理翻译人员除了应具备扎实的外语和母语基本功,还应熟悉相关专业的背景知识。一个对专业背景知识一无所知或知之甚少的译员只能依靠在翻译过程中对语言本身的理解去解释深刻的工程原理和工程现象,这对于译员来说是一件非常艰苦、枯燥,甚至痛苦的任务。在知识爆炸的现代社会,要求每个翻译人员掌握所有专业的工程知识是一种不切实际的苛求,但为了能胜任翻译的本职工作,能源电力工程管理翻译人员至少应熟悉当前正在参与的能源电力工程管理项目的基本知识。

翻译的过程不仅是语言运用和语义转换的过程,更是逻辑思维和逻辑分析的过程。针对专业背景知识的缺乏,译员在日常的翻译训练和翻译实践中,应注重专业背景知识的学习和积累,以丰富自己的知识储备,提高自己在相关专业领域的翻译能力。在每一次翻译实践中,译员应对文本中涉及的相关专业知识认真加以归纳和总结,把庞杂的专业知识一点一滴地积累起来。当然,每个人的能力和精力都是有限的,译员对于专业知识的学习,可以根据自身情况选择擅长的领域进行学习。翻译前,可以先向本专业人员了解相关知识,对原文有一定了解后再进行翻译,或者在翻译中遇到问题时,及时向专业人士请教,从而避免因专业知识的缺乏而造成翻译的失误。

此外,在能源电力工程管理英语中,很多相关的专业词汇和术语与普通词汇相同,但词义却完全不同,而且在不同的专业中词义也不一样;在不同领域,同一词汇会被译成意思完全不同的专业术语。在翻译过程中,有时甚至会遇到某词在词典上找不到适当的词义,如果任意地硬搬或逐字死译,会使译文生硬晦涩,不能确切地表达其原意,甚至造成误解。这时要根据上下文内容和逻辑关系,从该词的根本含义出发,结合相关专业去判断其词义。翻译是将一种相对陌生的表达方式转换成相对熟悉的表达方式的过程。其内容有语言、文字、图形、符号的翻译。其中,"翻"是指对交谈的语言转换,"译"是指对单向陈述的语言转换。工程技术翻译首先要忠实于原文,能准确传达出原文的信息和思想;其次,要通顺易懂、符合规范,还要具有专业特色;再次,语言要精练;最后,要做到对等,只有做到术语对等,才有可能实现信息对称,从而保证译文语言地道专业。

总之,对于能源电力工程管理翻译,首先要弄通文体所涉及的专业内容和有关的科学原理,不具备起码的基础知识和专门知识是不可能进行科学翻译的。其次要准确把握专业词汇的指称意义并注意固定词组的翻译,要区分一般用语和专业术语,然后再确定其专业范围。最后还要深入理解句子含义并灵活运用转换技法。翻译科技类文章不可拘泥字面,而要深入分析句子的命题结构、数量关系和逻辑关系,分清句子的主次和搭配的性质。

近年来,随着全球化进程的加速发展,科技类文本翻译的重要性日益凸显,那么如何培养和提高译者的专业素质呢?

第一,打好扎实的英汉双语基本功。

科技类文本由于内容、语域和语篇功能的特殊性,具有自身的一些特点。从事能源电力工程管理的翻译的译者既是原文的读者,也是译文的作者。作为读者,译者首先要通读原文,在通读原文之后,译者要具备分析原文语句的语法结构关系的基本功,特别是那些可能会影响对原文理解的结构复杂的句子。遇到修饰关系复杂的长句,我们可能很难很快弄清句子的意思,这时要集中注意力理顺句子中各个成分之间的语法关系,才能真正做到正确理解原文的含义。此外,在科技类英语文献中同一个词在不同的词组搭配中,在不同的语法结构中可能有不同的含义,同样需要译者具备词义辨析的能力。在深刻领会原文的基础上,重新进行信息结构整合,理清逻辑层次,用符合汉语习惯的表达方式将原文信息再现出来。可见,熟练地掌握英汉两种语言是从事科技翻译的基本条件。熟悉两种语言不仅要求译者具备两种语言的听说读写技能,同时还需掌握翻译所涉两种语言的语音、形态、语义,甚至包括句法和语用等方面的知识。此外,还应熟悉两种语言背后的文化,文化因素的差异可能会导致两种语言,甚至使用同一种语言的不同地区在专业词汇上的差异。

第二,积累中西方文化知识。

随着信息的传播和大众传媒的崛起,全球化与文化的关系更加紧密。翻译是一种跨文化信息交流活动,它的本质是传播。在全球化背景下,翻译的作用越来越凸显。简单来说,全球化使得语言与文化更加难以分开。语言是文化的载体,是传播文化的手段与途径;文化对语言有制约作用,发展并影响着语言。语言和文化是互相沉淀、相互辅助、流传而成的,语言其实也是文化的一部分,但是文化也依赖语言来进行传播。

无论从事哪个领域的翻译工作,译者所面临的不仅仅是语言的问题,还有两种语言背后的文化问题,翻译不仅是两种语言符号系统的转换,更是两种文化系统间的转换与交流。而从另一个角度看,语言反映一个民族的特征,它不仅包含着该民族的历史和文化元素,且蕴含着该民族对人生的态度和看法以及其特有的生活方式和思维方式。语言所反映的这种文化背景知识,是作者和源语读者所共有的,但是对于目的语读者来说是缺少的,译者这时候的主体性作用就要体现出来。译者必须作为桥梁去沟通两种语言与文化之间的断层,弥补源语和目的语之间的缺失,否则,译文就很难满足目的语受众的需要。

能源电力工程管理同样需要面对跨文化差异,需要考虑两种语言背后的社会历史发展差异。要想成功地翻译,不仅需要掌握两种语言,还要去了解两种文化。在具体文化背景当中,才可以将自身的意义再现出来。因此在翻译科技英语的时候,需要注重文化差异的影响,一些词语虽然看起来意思相同,但是涉及的文化含义却具有很大的差异。在实际翻译的时候,需要了解其文化内涵和使用语境,这样才能做到准确翻译。

第三,掌握相关领域的专业知识。

语言以其在社会生活中各个不同领域的使用而存在,使用在日常生活领域的是日常语言,使用在特殊领域的是特殊语言。随着全球化的发展,社会分工越来越精细,语言的使用也越来越专业化、精细化。译者要翻译特殊领域的语言就必须具备特殊领域的语言和行业知识。能源电力工程管理翻译工作者若缺少相关的专业知识,翻译出的译文则可能不准确、不专业。

第四,掌握必要的翻译技巧。

翻译实践表明,娴熟的翻译技巧对科技翻译工作者来说不可或缺。所谓翻译技巧,就是在弄清表达同一意义的外语和汉语异同的基础上,找出处理其不同之处的典型手法和转换规律,简言之,就是在处理源语词义、词序、句型和结构的时候所采用的手段。必要的翻译技巧不仅有助于弥合中外两种语言在表达方面的差异,为目的语的语言优势找到用武之地,而且更重要的是可以打造通顺地道的目标语理想译文。

可以说,掌握娴熟的翻译技巧是与外语水平、汉语功底和能源电力工程管理知识同等重要的基本素质,译文艺术之美,半数奠基于此。任何一种翻译技巧和方法都不可能凭空创造出来,而是在大量的翻译实践中积累、归纳和总结出来的。我们虽不能说翻译中有什么固定的"公式"可供译者临摹或类推,但灵活选择句式、巧妙调整词性和成分等跳出原文结构框框的技巧却是不容忽视的。掌握娴熟的翻译技巧有两种途径:一是勤于翻译实践,不断总结归纳;二是学习大家之长,借鉴他山之石。可见,掌握必备的翻译技巧是从事翻译工作的必要条件。在实际翻译过程中,译者只有灵活地使用各种翻译方法,译文才能地道准确、自然流畅。

3. 职业素养

在世界不同文明的交流与融合中,翻译始终起着不可或缺的重要作用。21世纪是全球化的世纪,是人类交流更加密切、交往更加广阔的世纪。随着全球化进程的加速,翻译作为沟通中外交流的桥梁和媒介,在让世界了解中国,让中国走向世界中发挥着不可替代的作用。中国五千年悠久而璀璨的历史文化不仅属于中国,也属于世界,中国理应对新世纪世界文化格局的形成和发展做出自己的贡献。而要承担和完成这一历史使命,中译外翻译工作任重而道远。翻译工作是决定对外传播效果的最直接因素和最基本条件,从某种角度来讲,也是一个国家对外交流水平和人文环境建设的具体体现;中译外在向世界说明中国、实现中外交流、介绍中国五千年文化、展示中华民族的追求和推动构建人类命运共同体中具有重要的作用。目前,中译外翻译工作面临的最大问题是高素质、专业化外译人才的严重匮乏和翻译队伍的"断层"。总之,翻译工作意义重大,任重而道远,因而对译者的职业素养要求也越来越高。良好的道德素养是翻译人员取得成功的先决条件,也是一名翻译工作者应尽的道德与法律义务。职业道德在不同的历史阶段具有不同的内容和含义。作为新世纪的翻译工作者,应具备以下基本的职业道德。

第一,遵守道德准则和底线。

译者在翻译过程中,应忠实于原文,不得擅自修改、增删,甚至伪造客户原件内容。译者还应讲诚信、重时效、保质量。如果所译稿件有时间要求,则应在规定的时间内保质保量地完成。只要接受了翻译业务,译者就有责任在整个翻译过程中全力以赴,认真翻译,以保证在承诺的时间内交稿。此外,译者还应保守秘密、遵守行规。应尊重译文使用者的合法利益,对因翻译需要而了解到的任何信息应视为职业秘密而守口如瓶。如果翻译内容为商业性文件、有价值的文件、专利技术等保密性材料,应对所译内容严格保密,既不能向第三方泄露也不可占为己有。同时,译者还应遵守行规,在工作中不应跟同行作不公平的竞争要价,故意割价抢走同行的客户;向客户收费必须合理,不得欺骗,不应接受低于行规或专业团体所规定的报酬。不可为多得稿费而增加字数使译文冗长,不得以译稿要挟客户等等。

第二,具有高度的社会责任感和良好的心理素质。

译者不得将所翻译的原件损坏、丢失,不得在原件上涂抹、翻译。译者不可以运用工作上获得的信息,牟取私利,比如译者不得利用原文所载资料买卖股票,否则可能触犯法律。此外,翻译工作者还需具备良好的心理素质。译者每天要面对众多的翻译文献,需要查阅大量的相关资料,这就对译员的心理素质提出了较为严格的要求。良好的心态、愉悦的工作心情,才能让译文更加客观,不带有个人感情色彩。

第三,具有严谨的翻译作风或翻译态度。

翻译无小事。翻译工作需要严谨的翻译作风或翻译态度,没有严谨的工作作风,抱着完成任务的态度去从事翻译工作,那么翻译出来的译文难免存在偏差,甚至还会造成利益损失。在具备丰富知识的同时,翻译工作者应始终保证一丝不苟的态度。翻译工作是一项科学性很强的工作,翻译者需要严谨、认真、准确,不断检查校对,以求精益求精。多数翻译中出现的错误都是由于翻译者的疏忽大意而造成的。由此可见,翻译者工作的态度决定着翻译质量和效果,影响着受众对所译内容的理解和认识。翻译者要不断培养自身科学的态度,严谨对待工作中的每一个词语。

第四,具有宽广的国际视野。

所谓国际视野,就是关注世界的现状与发展变化,了解主要文化及思维的基本特征,具有宽容理解、互利双赢的心态,具备追求人类和谐共处、共同进步的思想。所以,具有国际视野的翻译人才,应该做到以下几方面:(1)知晓中国,了解世界;(2)知晓中国的过去和现在,关注中国的未来;(3)了解世界的现状与变化,关心世界的发展;(4)把个人的发展和民族的复兴、国家的强盛和人类的进步紧密地结合在一起。这也是翻译人员,尤其是外宣翻译人员的基本职业素养。

众所周知,能源电力行业是关系到国计民生的基础性行业。能源电力工程管理翻译工作者肩负着重要的责任,在翻译的过程中务必要做到准确规范,因为任何的不准确或错误都会给科学研究、学术交流、生产发展带来不良影响乃至引发灾难,从而造成无可挽回的巨大损失。译者要不畏辛苦,勤学、勤练、勤译,还要时刻提醒自己要保持认真负责、一丝不苟、译风严谨的治学态度,不断提升自己的翻译水平。总的来说,从事翻译工作的人员应该具有强烈的责任心和道德感,具备良好的人格、品行和事业心,有职业目标、有责任感、讲诚信、懂感恩、知敬畏,也只有具备了这样的素质,才能读懂世界,读懂社会,读懂民生,成为一名具有崇高素养的翻译人才。

二、专业化与 IT 技术

信息时代向译者提出了掌握 IT 技术的要求,也给翻译产业带来了一场变革。现如今,信息量越来越大，传递速度越来越快。与之相适应的翻译量与日俱增，翻译速度成倍增长，翻译的无纸化程度急速提高。译者可利用网络与委托人沟通，通过网络试译、了解翻译要求、接收原文、传递译文。译者要利用委托人的数据库和自己的数据库在计算机上从事翻译工作，还要利用不同的翻译软件，特别是 CAT（Computer Aided Translation）软件与翻译记忆（Translation Memory）技术作为辅助翻译手段。从正面影响来看，翻译记忆技术可以减少翻译工作者的重复劳动，提升效率。从负面影响来看，采用翻译记忆技术进行翻译时，由于缺乏上下文，易造成误译，而一旦误译出现就可能被反复使用。无论是为了满足现代翻译项目的新需求，还是为了应对不断发展进步的机器翻译所带来的新挑战，翻译工作者都应在提高专业素质的同时，熟练掌握现代翻译技术，适应时代发展的需要。具体来讲，译者应具备以下几方面的翻译技术能力。

1. 计算机基本技能和 CAT 应用能力

计算机技术的基本应用能力已成为现代翻译职业人才的必备素质。在现代化的翻译项目中，翻译之前需要进行复杂文本的格式转换（如扫描文件转 Word）、可译资源抽取（如抽取 XML 中的文本）、术语提取、语料处理（如利用宏清除噪音）等，在翻译过程中需要了解 CAT 工具中标记的意义，掌握常见的网页代码，翻译之后通常需要对文档进行编译、排版和测试等等。可见，计算机相关知识与技能的高低直接影响翻译任务的进度和翻译质量。

此外，作为职业翻译人员，还需具备 CAT 工具使用能力，传统的翻译工作通常任务量不大，形式比较单一，时效要求也不是很强，所以并不强调 CAT 工具的作用。在信息化时代，翻译工作不仅数量巨大、形式各异，且突发任务多，时效性强，内容偏重商业实践，要求必须使用现代化的 CAT 工具。当前各大语言服务公司对翻译人员的招聘要求中，都强调熟练使用 CAT 或本地化工具。

2. 信息检索能力

在这个互联网高速发展的信息时代，信息广泛渗透到科技、文化、经济的各个学科领域以及人类生活的各个方面。18 世纪英国文豪、辞典编撰家塞缪尔·约翰逊曾指出，知识有两类，一类是我们自己知道的；另一类是我们知道在什么地方可以找到的。随着互联网的迅猛发展，网络上充斥着大量真假难辨的信息，快速、有效、经济地获取与自身需求相关的有用信息，已经成为信息化时代翻译人员一项不可或缺的基本技能。当代译员应熟练掌握主流搜索引擎和语料库的特点、诱导词的选择、检索语法的使用等，以提升检索速度和检索结果的质量。下面我们列举一些能源电力工程管理翻译可能用到的线上语料库检索平台、术语查询工具以及一些有用的网站。

1）通用语料库检索平台

北京外国语大学语料库检索平台：http://114.251.154.212/cqp 包含学术英语语料库、英语单语语料库、英语学习者语料库、平行语料库、翻译英语语料库、欧洲语言语料库、亚洲语言语料库、中文语料库等。

English-Corpora.org 语料库检索平台：https://www.english-corpora.org 收录世界上主要的英语语料库，包括英国国家语料库（British National Corpus）、美国当代英语语料库（Corpus of Contemporary American English）、全球网络英语语料库（Global Web-based English）等。

2）专用语料库检索平台

Tmxmall多功能能源电力语料检索平台：http://corpus.shiep.edu.cn/app/corpus/home

3）术语查询工具

中国特色话语对外翻译标准化术语库：http://210.72.20.108/index/index.jsp

联合国术语库：https://unterm.un.org/UNTERM/portal/welcome

术语在线：http://termonline.cn/index.html

中国规范术语：http://shuyu.cnki.net/index.aspx

4）常用网站

中国电力百科网：https://www.ceppedu.com/home/search-english.html

IEEE的数字资源平台IEEE Xplore：https://ieeexplore.ieee.org/Xplore/home.jsp

EI数据库：http://www.engineeringvillage.com/

需要指出的是，语料库和术语库等电子化资源虽然可以给翻译工作提供极大的便利，但它们始终只是翻译的辅助工具，无法完全保证所提供的资源准确无误。我们要学会恰当使用这些资源库，对检索查询到的材料进行合理判断；必要时通过不同的信息源进行多方互证，保证资料的可靠性。

3. 术语处理能力和译后编辑能力

所谓术语处理能力，即译者能够从事术语工作、利用术语工具解决翻译中术语问题的知识与技能。该项技能具有复合性、实践性强的特点，贯穿整个翻译流程，是翻译工作者不可或缺的一项职业能力。术语管理是译者术语能力的核心内容，已成为语言服务中必不可缺少的环节。译员可以通过术语管理系统（TMS）管理和维护翻译数据库，提升协作翻译的质量和速度，促进术语信息和知识的共享。因此，当代译员需具备系统化收集、描述处理、记录、存贮、呈现与查询术语管理的能力。机器翻译（MT）在信息化时代的语言服务行业中具有强大的应用潜力，与翻译记忆软件呈现出融合发展态势，几乎所有主流的CAT工具都可加载MT引擎。而随着AIGC时代的到来，生成式AI等新兴技术对语言行业也带来了新的变革，以Chat GPT为代表的技术也可以与CAT工具深度结合，高效打造语料库、优化术语库。智能化的机译系统可帮助译者从繁重的文字转换过程中解放出来，工作模式转为译后编辑。当代译员需要掌握译后编辑的基本规则、策略、方法、流程、工具等，这也是当代译员必备的职业能力。

以上是信息化时代职业译员翻译技术能力与专业化的主要构成要素。实质上，其中的每一项能力都与译者的信息素养密切相关。译者的信息素养指在翻译工作中，译者能够认识到如何快速准确获取翻译所需的信息，能够构建信息获取策略，使用各种信息技术工具检索、获取、理解、评判和利用信息，同时还要遵循信息使用的伦理要求。无论是上述哪一种技术能力，其本质都在于试图使用信息技术介入翻译过程，或是方便相关信息检索，或是自动化生成译文，或是对相关资源实施管理，以辅助译者将源语信息成功转化为目的语信息的过程，减轻译者的工作负担，提升翻译生产力。

第四章 能源电力工程管理英语翻译技巧

本章是基于前文所述翻译理论与技巧,有针对性地选用能源电力工程管理文本,从词汇翻译、句子翻译和语篇翻译三种视角的翻译方法进行分析与讲解。

第一节 词汇翻译视角

一、名词的译法

1. 直译法

- The center of an atom is called the nucleus. It is made of particles called protons and neutrons.

【译文】原子的中心称为原子核,原子核由质子和中子两种粒子构成。

- A generator can be broadly defined as a device that converts a form of energy into electricity.

【译文】广义上讲,发电机是把其他形式的能转化为电能的装置。

2. 转译法

1) 转译为动词

- Exposure to bright light may shorten storage life.

【译文】暴露于强光会缩短储存期。

【注解】原文中 exposure 本身为动词 expose 派生的名词,译文中根据汉语表达习惯转译为动词。

- The sequence of test is as follows: supply of materials and products; acceptance of materials; manufacturing of test specimens; application of process; inspection and testing of test specimens; provision of test reports.

【译文】试验程序如下所示:供应材料和产品;验收材料;制造试样;提交流程申请;试样检查和测试;提供试验报告。

【注解】本句列举了若干并列信息,英语中平行的名词化结构在译文中转译为多个动词短语。

2) 转译为形容词

❖ The power plant project involves a <u>variety</u> of tasks, including designing, construction, testing, and operation.

【译文】发电厂项目涉及各种各样的任务,包括设计、建设、测试和运营。

【注解】原文中variety本身为形容词various派生的名词,译文中根据汉语表达习惯转译为形容词。

❖ The optimization of energy consumption is of great <u>importance</u> for industry production.

【译文】能源消耗的优化对工业生产是非常重要的。

【注解】原文中的be of importance结构在句中作表语,译文根据汉语表达习惯,将名词importance转译为形容词"重要的"。

3) 转译为副词

❖ The team finds <u>difficulty</u> in managing the construction project within the allocated budget due to unexpected costs.

【译文】由于意外开销,团队<u>难以</u>在预算内管理施工项目。

❖ The project management software ensures <u>accessibility</u> to historical project data, facilitating insights for future project planning.

【译文】项目管理软件保证<u>方便</u>获取历史项目数据,以便从中获得对未来项目规划有益的洞见。

3. 增译法

❖ <u>Research</u> shows that the use of smart grids can optimize the distribution of electricity and reduce energy loss.

【译文】<u>研究结果</u>表明,使用智能电网可以优化电力分配并减少能源损失。

【注解】名词research后增译名词"结果"可以使译文更符合汉语的表达习惯。

❖ The electrical insulation in the equipment serves to protect operators from <u>electric shock</u>.

【译文】电气设备中的电绝缘体旨在保护操作员免遭<u>电击伤害</u>。

【注解】名词electric shock后增译名词"伤害"。

4. 省译法

❖ The elementary mechanical components of a machine are termed machine elements. <u>These elements</u> consist of three basic types: structural components, mechanisms, and control components.

【译文】机器的基本机械构件称为机器零件,包括三种基本类型,即:结构构件、机械构件和控制构件。

【注解】译文中省略了原文的these elements更加简洁清晰。

❖ Because of its simplicity, economy, and durability, the induction motor is more widely used for industrial <u>purposes</u> than any other type of a.c. motor, especially if a high-speed drive is desired.

【译文】与其他类型的交流电机相比,感应电动机因其简单、经济和耐用而被更广泛地用于工业,特别是需要高速驱动时更是如此。

【注解】译文中省译了原文的名词 purposes,更通顺流畅。

二、动词的译法

1. 直译法

- The project manager <u>allocates</u> resources to ensure timely completion of the project.

【译文】项目经理<u>分配</u>资源以确保项目及时完成。

- Water released from the reservoir flows through a turbine, spinning it, which in turn activates a generator to produce electricity.

【译文】从水库放出的水流过涡轮机,使其高速旋转,随后带动发电机产生电能。

2. 转译法

1) 转译为名词

- In science, it is important to <u>state</u> a law or a principle accurately.

【译文】在科学方面,定律或定理的准确叙述十分重要。

【注解】原文中的 state 在汉语译文中转译为名词"叙述",句尾修饰它的副词 accurately 也相应转译为形容词。

- The biomass energy system is chiefly <u>characterized</u> by its use of biological materials like plant matter or animal waste for power generation.

【译文】生物质能源系统的主要<u>特点</u>是使用植物物质或动物废物以产生能源。

【注解】原文的 characterized 在汉语译文中转译为名词"特点",修饰它的副词 chiefly 也相应转译为形容词。

2) 转译为副词

- Project engineers <u>tend</u> to conduct thorough feasibility studies before initiating a project to identify potential risks.

【译文】项目工程师<u>往往</u>在启动项目前会进行全面的可行性研究,以识别潜在的风险。

- The excessive use of air conditioners and heaters <u>boost</u> the increase of the power consumption.

【译文】过度使用空调和加热器<u>急剧地</u>增加了电能的消耗量。

【注解】原文中 boost 的宾语 increase 在汉语译文中转译为动词,boost 也相应由动词转译为副词"急剧地",从而使译文更加顺畅。

3. 省译法

- The task of ensuring a stable and efficient energy supply during peak demand times <u>has proved</u> especially demanding.

【译文】在需求高峰期确保稳定、高效的能源供应的任务特别繁重。

【注解】谓语动词大多是不可省译的,但是在有些语境中,比如本句的 has proved 在译文中被省略,更符合汉语的表达习惯。

- Evidently semiconductors <u>have</u> a lesser conducting capacity than metals.

【译文】显然,半导体的导电能力比金属差。

三、介词的译法

1. 直译法

❖ <u>With</u> the advancement of renewable energy technologies, the dependency on fossil fuels is gradually decreasing.

【译文】<u>随着</u>可再生能源技术的进步,对化石燃料的依赖正在逐渐减少。

❖ Proper maintenance is essential <u>to</u> prevent equipment failure.

【译文】适当的维护<u>对于</u>防止设备故障至关重要。

2. 转译法

1) 转译为动词

❖ The construction site requires workers <u>with</u> specialized skills in electrical engineering.

【译文】工程现场需要<u>具备</u>电气工程专业技能的工人。

【注解】介词转译为动词,因为英语中介词与及物动词在句法特征上具有共性,即二者都可以直接加宾语。另外很多介词来源于动词或者介词本身也带有动作意味,比如 across, around, along, beside, through, toward 和 with 等,可根据上下文需要转译为动词。

❖ The solar panels generate electricity <u>from</u> sunlight.

【译文】太阳能电池板<u>利用</u>太阳光发电。

【注解】相比介词直译"从太阳光发电",译文的动词"利用"更自然通顺。

2) 转译为连词

❖ <u>With</u> more active catalyst, the reaction is carried out completely.

【译文】<u>由于</u>使用了活性较大的催化剂,反应进行得很彻底。

【注解】原文中的介词 with 转译为连词后,逻辑关系更为清晰。

❖ The power station developed some new problems <u>on</u> its line-test.

【译文】发电站<u>在</u>试运行<u>时</u>出现了一些新问题。

3. 省译法

❖ The generator is connected <u>to</u> the grid <u>for</u> power distribution.

【译文】发电机与电网连接供电。

【注解】英语中使用介词较多,汉译表达时可根据意思省略,如本句中的介词 to 和 for。

❖ Whenever a current flows through a resistance, a potential difference exists <u>at</u> the two ends of the resistance.

【译文】电流通过电阻时,电阻的两端就有电势差。

【注解】省译介词 at。

四、代词的译法

1. 直译法

❖ <u>They</u> are regarded as electrons moving backward in time.

【译文】<u>它们</u>被认为是逆着时间而运动的负电子。

❖ This technology has the potential to revolutionize the power industry.

【译文】这项技术有可能彻底改变电力行业。

2. 转译法

1) 转译为名词

❖ <u>What</u> wind turbines do is harness the kinetic energy from wind and convert it into electrical power.

【译文】风力涡轮机的<u>作用</u>就是利用风的动能并将其转化为电能。

【注解】在这个句子中，what 是一个关系代词，引导了一个主语从句。what 指代并说明了风力涡轮机所做的事情，转译为名词"作用"。

❖ The electric resistance of copper is not so large as <u>that</u> of iron.

【译文】铜的电阻不像铁的<u>电阻</u>那样大。

【注解】指示代词 that 转译为名词"电阻"。

2) 转译为副词

❖ <u>All</u> of the heat lost by the warmer object in cooling is gained by the colder object.

【译文】温度较高的物体冷却时所失去的热量<u>都</u>被温度较低的物体吸收。

❖ <u>All</u> who have studied this question have come to the same conclusion.

【译文】研究了这一问题的人<u>都</u>得出了相同的结论。

3) 转译为连词

❖ AC can also be changed into DC by a device called a rectifier, <u>which</u> lets current flow only one way.

【译文】交流电也可以用一种叫整流器的装置转变成直流电，<u>因为该装置</u>只让电流单向流动。

【注解】原文中的关系代词 which 转译为连词后，逻辑关系更为清晰。

❖ The power plant implemented new safety measures <u>that</u> significantly reduced the risk of accidents.

【译文】发电厂采取了新的安全措施，<u>因此</u>事故风险显著降低了。

3. 省译法

❖ Solar storms can create beautiful auroras but they also have the potential to damage satellites and cause widespread power outages. Extreme solar storms battered Earth 1,000 years ago and if they strike again <u>they</u> could cause widespread blackouts.

【译文】太阳风暴会形成美丽的极光，但也会损坏卫星，引起大范围停电。极强的太阳风暴曾在一千年前冲击地球，如果再次来袭，可能导致大面积停电。

【注解】汉语代词比英语相对用得少，此处人称代词 they 根据汉语习惯省略不译。

❖ The solar panel, <u>which</u> converts sunlight into electricity, is an essential component of renewable energy systems.

【译文】太阳能电池板是可再生能源系统中将阳光转化为电能的重要组成部分。

【注解】关系代词 which 省略。

五、冠词的译法

1. 不定冠词 a/an 的译法

1）直译法

✧ The engineer installed a transformer to regulate the voltage in the power grid.

【译文】工程师安装了一个变压器来调节电网中的电压。

【注解】不定冠词 a 表示数量"一"。

✧ A power plant typically requires maintenance once a year to ensure its efficient operation.

【译文】一座发电厂通常每年都需要进行一次维护，以确保其高效运行。

【注解】不定冠词 a 放在某些名词前（如：day, week, year, month 等）之前，和名词一起作状语，译为"每一"或"一"。

2）省译法

✧ A transformer is an electrical device that converts a given input voltage into a different output voltage.

【译文】变压器是一种将给定输入电压转换为不同输出电压的电气设备。

【注解】不定冠词 a 在此例中表示事物类别，这种情况通常省略。

✧ Renewable energy is expected to play a key role in transitioning to a sustainable future.

【译文】可再生能源有望在实现可持续发展的过程中发挥关键作用。

【注解】不定冠词 a 省略。

2. 定冠词 the 的译法

1）直译法

✧ The project aims to reduce greenhouse gas emissions and promote clean energy alternatives.

【译文】该项目旨在减少温室气体排放，并推广清洁能源替代方案。

✧ The power plant is equipped with the latest technology for efficient electricity generation.

【译文】这座发电厂配备了最新的高效发电技术。

【注解】定冠词 the 起指示代词（this, that, these, those）的作用时，一般直译。

2）省译法

✧ The energy consumption of the industrial sector has a major impact on the environment.

【译文】工业部门的能源消耗对环境有重大影响。

✧ The implementation of energy-efficient technologies is crucial for reducing carbon emissions.

【译文】推广能效技术对于减少碳排放至关重要。

【注解】定冠词 the 在语法上起限定作用时通常省略。

六、连词的译法

1. 直译法

- The new energy-efficient appliances reduce electricity consumption, <u>but</u> they require an initial investment.

【译文】新型节能电器可以减少用电量,但需要一笔初期投资。

- <u>If</u> the grid system cannot handle the additional load, we need to consider upgrading the transmission lines.

【译文】如果电网系统无法承载额外负荷,我们需要考虑升级输电线路。

2. 省译法

- Electric charges, positive <u>and</u> negative, which are responsible for electrical force, can wipe one another out and disappear.

【译文】产生电场力的正负电荷会相互抵消掉。

【注解】省译了 and 更符合汉语的表达习惯,即汉语可以把两个或更多的词语或句子连起来而不用连词。

- The truth is <u>that</u> the current increases with every decrease of resistance.

【译文】事实是,电流随每次电阻减小而增加。

【注解】本例中 that 只是作为表语从句的引导词而不具有特定意思,因此省去不译。

3. 增译法

- <u>Although</u> solar power is a clean and renewable energy source, its efficiency is still limited by weather conditions.

【译文】虽然太阳能是一种清洁且可再生的能源,但它的效率仍受制于天气条件。

- <u>Because</u> this current continually reverses in direction, it is called an alternating current.

【译文】由于这种电流不断地改变方向,故称为交流电。

【注解】英语的从句中通常只需要一个连词连接主句,而汉语的连词往往成对出现,如 because 在译成汉语时,很可能会译成"因为……所以",although 会译成"虽然……但是"等。在此种情况下,应作适当的补充以符合汉语的行文习惯。

七、形容词的译法

1. 直译法

- The <u>advanced</u> grid integration technology enables seamless integration of distributed energy resources into the power grid.

【译文】先进的电网集成技术使分布式能源资源无缝接入电力网络。

- The project requires a <u>comprehensive</u> risk assessment to identify potential hazards and develop appropriate mitigation strategies.

【译文】该项目需要进行全面的风险评估,以确定潜在风险并制定适当的缓解策略。

2. 转译法

1) 转译成名词

❖ Metals in general have high electrical conductivity, high thermal conductivity, and high density. Typically they are malleable and ductile, deforming under stress without cleaving.

【译文】一般情况下,金属具有高导电性、高导热性和高密度。它们的典型特征是具有可锻性和可延展性,在重力作用下会产生变形但不会裂开。

【注解】有些表示物质属性的形容词如 ductile, malleable, combustible, compact, comparable, compatible, elastic, fluid, kinetic, plastic, static 等作表语时,往往转译成名词。原文中的形容词 malleable 和 ductile 转译为名词,和第一句的"高导电性""高导热性"和"高密度"也形成自然呼应。

❖ Conflict resolution and team collaboration are by no means less important than technical expertise in efficient project management.

【译文】在高效的项目管理中,冲突解决和团队协作的重要性绝不亚于技术专长。

【注解】在 as+形容词+as 或形容词比较级+than 句型中,其中的形容词经常可以译为名词。

2) 转译为动词

❖ It's unnecessary to be anxious about the project budget, as we have implemented robust cost control measures and regularly monitor expenses to ensure that they remain within the allocated budget.

【译文】不必担心项目预算,因为我们已经实施了强有力的成本控制措施,并定期监控开支,确保其保持在预算范围之内。

【注解】和系动词连用的表示心理活动或心理状态的形容词作表语时,通常可以转译为动词。比如:sure, careful, cautious, familiar 等。

❖ Removed parts shall be cleaned and the cleaned surface must be free from deposits and all traces.

【译文】零件拆除后,应进行清洗,清洁后的表面不得存在沉积物或任何痕迹。

【注解】free 本身具有动词意义,因此转译为汉语的动词也比较自然。

3) 转译为副词

❖ All of this proves that we must have a profound study of the economic feasibility and environmental impact of transitioning to renewable energy sources.

【译文】这一切都证明了我们必须深入地研究向可再生能源转型的经济可行性和环境影响。

【注解】本句形容词 profound 修饰的名词 study 转译为动词,相应地 profound 也转译为副词。

❖ Robust risk management practices and contingency planning are an absolute necessity in power plant construction projects.

【译文】强大的风险管理实践和应急计划在电厂建设项目中是绝对必要的。

【注解】本句形容词 absolute 修饰的名词 necessity 转译为形容词,相应地 absolute 也转译为副词。

3. 增译法

❖ China's load centers and coal resource centers are <u>separated</u>, which poses challenges for the transmission of electricity.

【译文】中国的负荷中心和储煤中心<u>分布在不同地区</u>,这对电力传输构成了挑战。

【注解】原文中的形容词 separated 在译文中进行了增补说明,使得译文意思更加完整清晰。

❖ These companies seemed to be successful in every aspect of the competitive model: they were fast to market and <u>efficient,</u> cost-conscious, and customer-focused.

【译文】这些公司似乎在竞争模式的每个方面都取得了成功;他们快速进入市场并<u>高效运作</u>,注重成本,关注客户。

【注解】本句中三个形容词并列,由于后两个形容词 cost-conscious 和 customer-focused 转译为四个字的动词,因此第一个形容词 efficient 在转译为动词的时候也根据后两个形容词的译法进行了增补。

八、副词的译法

1. 直译法

❖ Electrical engineers need to <u>carefully</u> integrate different power generation sources to ensure the stability and reliability of the electricity supply.

【译文】电气工程师需要<u>仔细</u>整合不同的发电源以确保电力供应的稳定性和可靠性。

❖ With the increase in people's environmental awareness and the professional needs of some outdoor sports enthusiasts, the market for photovoltaic application products has grown <u>significantly</u> in recent years.

【译文】随着人们的环保意识提高和一些户外运动爱好者的专业需求,近年来光伏应用产品市场<u>大幅</u>增长。

2. 转译法

1) 转译为名词

❖ The capacity of the power grid is not capable of handling peak loads, this is <u>why</u> we need to upgrade our infrastructure.

【译文】电网的容量无法处理高峰负载,这就是我们需要升级基础设施的<u>原因</u>。

【注解】表示原因的副词 why 译成汉语时转变为名词"原因"。

❖ Supported by advanced technology and innovative solutions, electricity has been <u>successfully</u> generated by geothermal resources in our recent project.

【译文】在我们最近的项目中,得益于先进的技术和创新性的解决方案,地热资源发电获得了<u>成功</u>。

【注解】副词 successfully 转变为汉语中作宾语的名词"成功"。

2) 转译为动词

❖ <u>Admittedly</u>, the initial investment for renewable energy projects can be high, yet the long-term financial and environmental benefits are well worth it.

【译文】不可否认,可再生能源项目的初期投资可能会很高,然而从长期来看,其财务和环保收益是非常值得的。

❖ With the switch off, the power supply to the entire building was interrupted.

【译文】开关关闭后,整栋建筑的电力供应被中断。

3) 转译为形容词

❖ To detect gases, a sample of air can be analyzed chemically.

【译文】可以通过对空气样本的化学分析来探测气体。

【注解】原文中的动词 analyze 转译成汉语的名词,因此修饰 analyze 的副词 chemically 也转译成汉语的形容词。

❖ The project management approach is chiefly featured by its focus on risk assessment, team synergy, and continuous performance evaluation.

【译文】这种项目管理方法的主要特点是重视风险评估、团队协同作用和持续的绩效评估。

九、数词的译法

1. 数字的译法

1) 直译法

2 066 km 译为 2 600 千米

200 v 译为 200 伏

50 t 译为 50 吨

2) 转换计量单位

9,600,000 square kilometers 译为 960 万平方千米

53,426,000 cubic meters 译为 5 342.6 万立方米

300 KV 译为 30 万伏

3 000 MW 译为 300 万千瓦

【注解】表示大数时,欧美各国采用千进制,即以 kilo-(千)或 mil-(千的二次方,即百万)为计数单位。但中文习惯以万为计数单位的万进制。因此英文中的千进制,译为汉语时为万进制更符合汉语的表达。

2. 倍数增减的译法

1) 倍数增加的译法

❖ If carefully maintained, this machine will have its working life increased by 5 times.

【译文】如果仔细保养,这台机器的使用寿命将会延长至原来的 5 倍。(延长了 4 倍)

【注解】英语中的 increase(by) n times,是包括基数在内,表示增加后的实际结果,意思是"增加到几倍";而汉语中的"增加了多少倍",只表示增加的倍数,不包括基数。因此,英译汉时,要将原来的倍数减去 1,译为"增加了 n-1 倍"或者"增加到 n 倍"。

常用倍数增加表达法及译法有下列 3 种:

increase by n times 译为"增加了 n-1 倍"或"增加到 n 倍"

increase n times　译为"增加了 n-1 倍"或"增加到 n 倍"
increase by a factor of n　译为"增加了 n-1 倍"或"增加到 n 倍"

❖ The speed exceeded the average speed by a factor of 4.5.

【译文】该速度超过平均速度 3.5 倍。(或"该速度是平均速度的 4.5 倍")

2) 倍数减少的译法

❖ The automatic assembly line can shorten the assembling period(by)ten times.

【译文】自动装配线能够将装配时间缩短为原来的 1/10。(或降低 9/10)

【注解】常用的倍数减少表达法及译法有下列 3 种：

decrease by 3 times　译为"减至 1/3(减少了 2/3)"
decrease 3 times　译为"减至 1/3(减少了 2/3)"
decrease by a factor of 3　译为"减至 1/3(减少了 2/3)"

❖ The equipment under research is likely to reduce the error possibility by a factor of 3.

【译文】正在研制的设备有可能会把误差率降低至原来的 1/3。(或"降低 2/3")

3) 倍数比较的译法

❖ This metal is 3 times heavier than that one.

【译文】这种金属的重量是那种金属的三倍。

【注解】常用的倍数增加比较结构表达法及翻译方法如下：

A is n times larger than B 译为"A 比 B 大 n 倍"
A is n times as large as B 译为"A 是 B 的 n 倍"
A is n times B 译为"A 是 B 的 n 倍"
A is larger than B by n times 译为"A 比 B 大 n 倍"

❖ The wire is 3 times shorter than that one.

【译文】这根导线只有那根导线 1/3 长。

【注解】A is 3 times smaller than B 译为"A 是 B 的 1/3"或"A 比 B 小 2/3"。

第二节　句子翻译视角

　　能源电力工程管理英文文本的一个重要特点是广泛使用复合句。复合句可能由一个主句和一个或多个从句组成。了解并掌握不同类型从句的译法对理解和翻译文本非常重要。英语的从句大致可分为名词性从句(主语从句、宾语从句、表语从句和同位语从句)、形容词性从句(定语从句)和副词性从句(状语从句)。本节重点讲解并列句以及各类从句的具体翻译方法，最后再对复杂长句的翻译方法进行介绍。

一、并列句的译法

　　英语的并列句是由两个或更多的简单句组成，这些简单句通常由并列连词，如 and, or, but 等连接。因为每个简单句都能独立地表达一个完整的想法，并列句通常直译或者被转译为复合句。

1. 直译
- Solar energy is unlimited in supply, but its exploitation and utilization are limited owing to the limitation of technology and conditions.

【译文】虽然太阳能是无限的,但由于技术和条件的限制,其开发利用是有限的。

【注解】本句采取直译法,即分别翻译并列连词 but 前后的简单句中,并将 but 直译。

- The budget for the project has been approved, and the team is now focusing on recruiting the necessary personnel.

【译文】项目的预算已经获得批准,团队现在正在专注于招聘必要的人员。

【注解】本句中省译了并列连词 and。

2. 译成复合句
- Implement energy-saving measures and the power consumption of the factory will significantly reduce.

【译文】如果实施节能措施,工厂的电力消耗将会显著降低。

【注解】当 and 连接的两个分句中,前一个分句是祈使句时,祈使句往往表示条件的含义,汉译时可以译为条件从句,可增译"如果""要是"等。

- The project manager is adjusting the budget allocation, for they need to accommodate for unexpected costs that may arise during execution.

【译文】项目经理正在调整预算分配,因为他们需要应对执行过程中可能出现的意外成本。

【注解】for 作为并列连词时常常译为"因为",但语气不强,只是对前一个分词作出补充性说明。

二、主语从句的译法

主语从句在复合句中位于主语位置,一般在谓语之前:主语从句+谓语+其他成分;还有 it 作形式主语,主语从句放在谓语后的情况:it+谓语+主语从句。

1. 顺译法
- Whether we can achieve our goal of transitioning to 100% renewable energy by 2050 depends on the advances in power storage technology.

【译文】我们能否在2050年之前实现向100%可再生能源过渡的目标,取决于储能技术的进步。

【注解】本句的结构为"主语从句+谓语+其他成分",这种情况往往采用顺译法,即把主语从句译在句首。

- It is certain that improving energy efficiency in buildings can significantly reduce electricity consumption.

【译文】可以肯定,提高建筑能效可以显著减少电力消耗。

【注解】本句的结构为"it+谓语+主语从句",即以 it 作形式主语引出真正的主语从句。按照原文中句子的词序,先译主句(it 省译),然后再译主语从句。这句话同时也使用了分译法,也就是把 It is certain 拿出来单独翻译,突出了主句的语气。

2. 倒译法

❖ It is also important that consumers are educated about energy conservation to reduce power wastage.

【译文】教导消费者如何节能以减少电力浪费也是非常重要的。

【注解】本句的结构为"it + 谓语 + 主语从句",根据汉语表达顺序,译文采取倒译法,即先把主语从句译出来,再译主句(it 省译)。

❖ It matters much to efficiency that the tasks are distributed according to the team members' expertise.

【译文】根据团队成员的专业知识分配任务,这对效率非常重要。

【注解】本句的结构为"it + 谓语 + 主语从句",译文采取倒译法,先译主语从句。同时因为从句较长,这句话也使用了分译法,也就是把 It matters much to efficiency 拿出来单独翻译,把 it 译出来强调对效率的重要性,从而使译文重点突出,结构分明。

三、宾语从句的译法

宾语从句在复合句中作宾语。宾语从句分为两类:一是动词后面的宾语从句,包括不定式、分词、动名词后的宾语从句;二是介词后面的宾语从句。翻译宾语从句大都采取顺译法,有时也会考虑汉语的表达习惯采取倒译法。

1. 顺译法

❖ This sourcebook illustrates how PV modules can be designed as aesthetically integrated building components (such as awnings) and as entire structures (such as bus shelters).

【译文】这本参考资料阐述了如何将光伏组件设计成美观的集成建筑构件(如遮阳篷),或是整个建筑结构(如公共汽车候车亭)。

【注解】原文为"主谓宾"结构,宾语从句在动词后面。由于主谓宾结构的句型也是汉语常用的句型,因此译文直接用顺译法按照英文原文顺序译出。

❖ The efficiency of a solar panel does not depend on where you install it, but on how much sunlight it can receive.

【译文】太阳能板的效率并不取决于你在哪里安装它,而取决于它能够接收到多少阳光。

【注解】本句为"主谓宾"结构,宾语从句在介词 on 后面,语序和中文一致,因此顺译。

2. 倒译法

❖ The engineering team did an excellent job, except that they underestimated the cost for renewable energy integration.

【译文】除了低估了可再生能源集成的成本,工程团队做得很好。

【注解】原文为"主谓宾 + 状语"结构,主干部分放在前面,后面的状语部分由介词 except 加定语从句构成。由于汉语习惯将状语传达的背景信息放在前面,因此译文采用倒译法,将 except 后面的定语从句 that they underestimated the cost for renewable energy integration 先译出。

❖ Field testing will show whether the newly developed battery can store enough solar power.

【译文】新开发的电池能否储存足够的太阳能,实地测试就能证明。

【注解】相对英文固定的主谓宾结构,译文的句子更加灵活,通过倒译法突出了宾语从句想表达的内容。这句话同时也使用了分译法,也就是把 Field testing will show 拿出来单独翻译,突出了主句的语气,整个句子也更加清晰明了。

四、表语从句的译法

表语从句在复合句中作表语,相当于名词,位于系动词之后。一般采用顺译法,偶尔采用倒译法。

1. 顺译法

- The magnetic force on current-carrying wires is <u>what makes electric motors work</u>.

【译文】作用于载流导线的磁力就是<u>使电动机工作的力</u>。

【注解】原文为"主系表"结构,与汉语的构句习惯比较接近,因此采用顺译法。

- The challenge is <u>that shifting to renewable energy sources requires significant initial investments</u>.

【译文】挑战在于,<u>转向可再生能源需要大笔的初期投资</u>。

【注解】原文为"主系表"结构,采用顺译法。这句话同时也使用了分译法,也就是把 The challenge is 拿出来单独翻译,突出了主句的语气。

2. 倒译法

- The consequence of improper risk assessment is <u>that potential hazards in a power project may go unnoticed until it's too late</u>.

【译文】<u>电力项目中潜在的危险可能在事态发展到无法挽回时才被察觉</u>,这就是风险评估不当的结果。

【注解】由于原文表语从句较长,如果按顺译法译为"风险评估不当的结果是在电力项目中潜在危险可能在事态发展到无法挽回时才被察觉",译文较为冗长,所以此处用倒译法,先译从句,再译主句。同时采取分译法,在从句和主句之间用"这就是"连接,突出原文的因果关系。

- That's why <u>a comprehensive environmental impact assessment is crucial before initiating any energy infrastructure project</u>.

【译文】<u>在启动任何能源基础设施项目之前,进行全面的环境影响评估是至关重要的</u>,其原因就在这里。

【注解】由于原文表语从句较长,如果按顺译法译为"这就是为什么在启动任何能源基础设施项目之前进行全面的环境影响评估是至关重要的",译文较为冗长,所以此处用倒译法,先译从句,再译主句。That's why 可以译为"原因就在这里""理由就在这里"或者"道理就在这里"等。

五、同位语从句的译法

同位语从句是用来给名词或代词作进一步解释或说明的从句。它通常紧跟在名词或代

词后面,并且用来表达这个名词或代词的含义、定义、解释、特点等。同位语从句的译法主要有以下几种:

1. 顺译法

- He expressed the hope <u>that renewable energy sources would be prioritized in future power generation plans.</u>

【译文】他表示希望<u>在未来的发电计划中优先考虑可再生能源</u>。

- They came to the conclusion <u>that not all things can be done by automation.</u>

【译文】他们得出结论:<u>并非所有的事情都能由自动化来完成</u>。

【注解】译文采取顺译法,保持原句结构和意思不变。其中同位语从句译成了一个独立的句子。

2. 倒译法

- The belief <u>that a diversified energy mix is key to ensuring a stable power supply</u> has guided energy policy decisions.

【译文】<u>多样化的能源结构对确保稳定的电力供应至关重要</u>,这一信念指导着能源政策的决策。

【注解】原文中的同位语从句 that a diversified energy mix is key to ensuring a stable power supply 位于主语 the belief 和谓语 has guided 之间,此时使用倒译法,先把从句译成独立的句子放在句首,然后再译主句。同时由于从句比较长,在主句前加译"这一"连接从句和主句,使得译文层次更加清晰。

- The fact <u>that power demand is increasing at an alarming rate</u> necessitates the expansion of electricity generation capacity.

【译文】<u>电力需求以惊人的速度增长</u>,这一事实要求扩大发电容量。

3. 转译法

- The thought suddenly came to the engineer's mind <u>that the problem would be moderated to some degree if he added a few more components to the device.</u>

【译文】工程师突然想到,<u>如果他在装置上再加几个组件,问题就可以得到某种程度的缓解</u>。

【注解】原文中的 thought 本身就是含有动词意义的名词,译文将其转译成动词,其同位语从句 that the problem would be moderated to some degree if he added a few more components to the device 也转译为谓语动词的宾语,相当于英语的宾语从句。

- These uses are based on the fact <u>that hydroelectric power plants utilize the gravitational force of flowing water to generate electricity.</u>

【译文】这些用途是基于<u>水电厂利用流水的重力来发电</u>的事实。

【注解】原文虽然是同位语从句,但是意义上和形式上接近定语从句,因此译文把同位语从句提前,转译为定语。

- <u>In spite of the fact that the project was delayed due to unforeseen circumstances,</u> the team successfully completed the power plant construction within the revised timeline.

【译文】<u>尽管由于意外情况导致项目延迟</u>,但团队成功地在修订的时间表内完成了电建设。

【注解】根据原文主从句的逻辑关系，that 引导的同位语从句可以译为表示让步的状语从句，"介词 + fact + that"结构可以看作是复合的从属连词。

六、定语从句的译法

定语从句是由关系代词或关系副词引导的从句，其作用是作定语修饰主句的某个名词性成分，一般紧跟在它所修饰的先行词后面。

定语从句可分为限定性定语从句、非限定性定语从句和状语化从句等，翻译时可采用合译法、分译法和转译法。

1. 合译法

❖ The negative smoke particles are then attracted to the inside surface of the surrounding tube which is positive.

【译文】然后，带负电的烟粒子被吸向周围带正电的管子的内侧面。

【注解】原文中 which 引导的限制性定语从句结构比较简单，采取合并的方法将后置定语从句译成"……的"，置于所修饰的词语 tube 前面，从而将复合句译为汉语单句。

❖ Biofuels critics say current US ethanol, which is mostly made from corn, does little to reduce greenhouse gas emissions.

【译文】生物能源批评人士指出，目前美国以玉米为原料生产出来的乙醇并没有对温室气体减排做出多大贡献。

【注解】非限制定语从句最明显的标志是与被它所修饰的词用逗号分开。原文中 which 引导的非限制性定语从句比较简短，和主句关系也比较密切，在译文中将定语从句翻译成"……的"，置于所修饰的词语 ethanol 前面。

2. 分译法

❖ The engineer worked out a new method by which wind turbines can efficiently harness and convert wind energy into electricity.

【译文】工程师设计了一种新的方法，通过这种方法，风力涡轮机可以高效地利用并将风能转化为电力。

【注解】原句中 by which 引导的限制性定语从句比较长，因此将定语从句和主语分开来译，使得译文结构清晰明了。

❖ The concept of energy leads to the principle of the conservation of energy, which unifies a wide range of phenomena in the physical science.

【译文】能量这一概念引出了能量守恒定律，该定律统一了物理科学中相当多的现象。

【注解】原句中 which 引导的非限制性定语从句是为了补充说明前面的 principle of the conservation of energy，在译文中采取分译法，即将从句译为并列分句，分句主语用该定律代替先行词。

3. 转译法

❖ An electrical current begins to flow through a coil, which is connected across a charged capacitor.

【译文】如果把线圈跨接到已充电的电容器上，电流就开始流过线圈。

【注解】定语从句在特定的上下文可以有不同的逻辑含义。原文由 which 引导的从句虽然在语法功能上属于非限制性定语从句,但是从逻辑关系上分析,又具有状语功能,因此在译文中,将其转译成与汉语对应的表示条件的状语从句。

❖ The newly invented robots, which can do various kinds of work for human beings, cannot replace human beings in all aspects.

【译文】新型的机器人尽管能够做各种各样的工作,但是终究不能在各个方面取代人类。

【注解】将定语从句转译成表示让步的状语从句。

4. 融合法

❖ There are many people who are interested in this new invention.

【译文】很多人对这项新发明感兴趣。

【注解】在限制性定语从句,尤其是"there be"句型中,可以把主句和定语从句融合成一个简单句,从句谓语译成主句谓语。

❖ There are some metals which possess the power to conduct electricity and the ability to be magnetized.

【译文】有些金属具有导电能力和磁化能力。

【注解】原文中的主句 there are some metals 被压缩成一个短语作主语,然后与 which 引导的限制性定语从句融合在一起,译成一个独立句。

七、状语从句的译法

英语的状语从句一般由从属连接词和起连词作用的词组来引导。按照句法功能划分,大致可以分成时间状语从句、地点状语从句、条件状语从句、原因状语从句、目的状语从句、让步状语从句、结果状语从句、方式状语从句和比较状语从句。

1. 时间状语从句的译法

❖ When water is in short supply, it is desirable to use the limited available potential energy sparingly, for periods of short duration, to meet the peak-load demands.

【译文】在供水量不足时,人们希望能够短期地节约使用可获得的有限的潜在能量,以满足负荷的高峰需求。

【注解】本句采用了直译和顺译法。原文主句是 it is desirable to...,其中 it 是形式主语,动词不定式短语是真正的主语。原文 When 引导的时间状语从句和汉语习惯一致,都是放在句首,因此采取顺译法。

❖ The overall efficiency is only about 65% to 70%, but the economics are frequently favorable when one considers the economics of the overall system, including the thermal units.

【译文】其整体效率(指发电机当作电动机运行方式下的)仅为65%～70%,但当我们考虑到包括火力发电机组在内的整个系统的经济指标时,其效益往往还是相当令人满意的。

【注解】本句采用了直译和倒译法。but 引导的并列句中,有一个 when 引导的状语从句,翻译时按汉语的习惯,置于并列句的句首。

2. 地点状语从句的译法

❖ In regions <u>where there is high wind speed</u>, wind farms are more feasible for power generation.

【译文】<u>在风速较高的地区</u>,风电场更适合发电。

【注解】本句采用了<u>直译和顺译法</u>。

❖ No relief valve is required <u>where two pressure reducing valves are installed.</u>

【译文】如果装上两个减压阀,就不需要安全阀。

【注解】本句采取了倒译法和转译法。原文虽然从形式上看是 where 引导的地点状语从句,但从逻辑上看有条件状语的意义,因此译文中转译为条件从句。因为条件从句按照汉语习惯一般放在主句之前,所以本句用了倒译法。

3. 条件状语从句的译法

❖ If the countries currently exceeding this level could decrease their energy consumption to within this range, their quality of life would be maintained, and global resources would be better preserved and utilized。

【译文】如果超过这个范围的国家能够将能源消费降低至该范围内,它们的生活质量将得到保持,并且全球资源将得到更好的保护和利用。

【注解】本句中条件状语从句和汉语习惯一样位于主句之前,因此译文采取了顺译法,将 if 引导的条件状语从句放在句首。

❖ No one is allowed to carry out any construction activities at the site <u>unless they have obtained a special permit.</u>

【译文】<u>除非获得特别许可证</u>,任何人都不允许在该地点进行任何建设活动。

【注解】本句译文按照汉语习惯将 unless 引导的条件状语从句放在主句之前进行倒译。但是根据上下文如果特别强调主句的语气,也可以按与原文顺序译为"任何人都不允许在该地点进行任何建设活动,除非获得特别许可证。"这时条件状语从句起到了补充说明的作用,被转译为汉语的补语从句。

4. 原因状语从句的译法

❖ The energy is not "free" <u>since the capital investment in dams, transmission, and generation must be accounted for.</u>

【译文】<u>由于必须将大坝、输电及发电设备上的资金投入计算在内</u>,所以水力发电并非无需花费。

【注解】按照汉语习惯,原因状语从句一般放在主语从句前,因此译本采取了倒译法,将 since 引导的原因状语从句译在句首。

❖ Higher global temperatures would probably cause glaciers to melt, as well as leading to an expansion of the oceans <u>because warm water occupies a larger volume of space than cold water.</u>

【译文】全球气温的上升很可能将使冰川融化,并将导致海洋扩张,<u>因为温水所占的体积比冷水大</u>。

【注解】原文 because 引导的原因状语从句虽然在主句后面,但是考虑上下文的重点在主句,从句主要起到补充说明的作用,因此译文仍采用顺译法,将从句保留在后面。

5. 目的状语从句的译法

❖ An atomic furnace must be surrounded by heavy thick walls <u>so that the rays cannot get out</u>.

【译文】<u>为了防止射线漏出</u>，必须用很厚的墙壁将原子反应堆屏蔽起来。

【注解】原文中 so that 引导的目的状语从句译作汉语的"为了"，这种情况通常根据汉语习惯将状语从句放在句首。如果将 so that 译为"以防"，状语从句也可以放在句尾，译为"必须用很厚的墙壁将原子反应堆屏蔽起来，以防射线漏出"。总之，译文中的状语从句的位置可以在句首（译作"为了""要"等），也可以在句末（译作"以便""为的是""以免""以防"等），要根据上下文的意思灵活处理。

❖ Cables are usually laid underground <u>so that their life may be prolonged</u>.

【译文】电缆通常铺设在地下，<u>以便延长使用寿命</u>。

【注解】原文中 so that 引导的目的状语从句译作汉语的"以便"，在译文中习惯上将这种目的状语从句放在句末。

6. 结果状语从句的译法

❖ Gears play such an important part in machines <u>that they have become the symbol for machinery</u>.

【译文】齿轮在机器中起着相当重要的作用，<u>以至于它已经成为机械的象征</u>。

【注解】结果状语从句无论在英语和汉语中，都习惯放在主句之后，因此多采取顺译法。

❖ This generator is so outdated <u>that it no longer meets the energy efficiency standards</u>.

【译文】这台发电机太过时了，<u>已经不再符合能效标准</u>。

【注解】顺译法。

7. 让步状语从句的译法

❖ Distribution substations are typically the points of voltage regulation, <u>although on long distribution circuits (of several miles/kilometers), voltage regulation equipment may also be installed along the line</u>.

【译文】<u>尽管在较长的配电线路（几英里/公里）上，也可以沿线安装电压调节设备</u>，但配变电站是更典型的电压调节点。

【注解】此句中，although 引导陈述语气的让步状语从句，译为尽管。根据汉语习惯，although 引导的让步状语从句位于句首。除了直译从属连词之外，还要使用成对的关联词"虽然……但是"。

❖ <u>Although derived from the sun in a way which is not fully understood</u>, heat is thought to be with invisible electro-magnetic waves, which are able to pass through a vacuum.

【译文】<u>虽然人们还不完全了解热能是如何来自于太阳的</u>，但他们认为热能带有一种看不见的并能穿越真空的电磁波。

【注解】本句中，although 引导的让步状语从句里 although 后面省略了主语和谓语。由于从句的主语和主句的主语指代的是相同的事物，可以判断 although 后面省略了 heat is。在译文中将省略的部分译出。同时根据汉语习惯将原文的被动语态译为主动语态。

8. 方式状语从句的译法

❖ Electricity flows through a wire just as water flows through a pipe.

【译文】电流通过导体，正像水流过管子一样。

【注解】本句采用顺译法。

❖ Solid expand and contract as liquids and gases do.

【译文】固体像液体和气体一样膨胀和收缩。

【注解】本句采用合译法，即把原文的主句和 as 引导的方式状语从句译为一个单句。

9. 比较状语从句的译法

❖ The broader the range of network services is, the more capabilities a network can provide to users connected to it.

【译文】网络服务的范围越广，网络就能为接入的用户提供越多的功能。

【注解】本句采用顺译法。

❖ Electronic engineers are more interested in the application than in the development of semiconductor devices.

【译文】电子工程师对半导体器件的应用比对半导体器件的研究更感兴趣。

【注解】原文的主句 in the application 后面省略了 of semiconductor devices。在译文中要将省略的部分译出。

八、复杂长句的译法

能源电力工程项目管理英语因为涉及能源电力科技等专业知识，表达中长句比较多。英语长句的特点是语言结构层次多而复杂：句子中并列成分多，各种短语、修饰语多，并列分句或各种从句多。由于这些特点，我们在进行长句翻译时，第一步要厘清原文的句法结构，通过分析句子内部每一个单独的主谓关系之间的主与次、修饰与被修饰等关系，把握住这句话的中心思想。第二步，根据汉语的表述风格和习惯，调整句子语序，此时可以运用前面几章所说的各类词语和复合句等的翻译技巧对每个句子成分进行初步翻译。第三步回读整个译文，根据上下文增补潜在信息，力求使译文准确通顺。

翻译英语长句，主要可采用以下四种方法。

1. 顺译法

❖ As the products of power enterprises, which have the features of production, supply, sales and meanwhile keep balance with time, are electric energy and the stoppage of power production will cause great losses to the national economy, the basic task of power enterprises is to provide sufficient, reliable, qualified and cheap electric energy for users, and meanwhile to accumulate abundant capitals for the state and to promote the development of national economy and to improve people's lives.

【分析】第一步，厘清原文的句法结构。

本句为复合句，主要由主句和状语从句组成。其具体句法结构依次是：

① As the products of power enterprises, which have the features of production, supply, sales and meanwhile keep balance with time, are electric energy and the stoppage of power production will cause great losses to the national economy 为原因状语从句,其中包括画线部分 which 引导的非限制性定语从句。

② the basic task of power enterprises is <u>to</u> provide sufficient, reliable, qualified and cheap electric energy for users, and meanwhile <u>to</u> accumulate abundant capitals for the state and <u>to</u> promote the development of national economy and <u>to</u> improve people's lives 为主句,主语为 the basic task of power enterprises,表语是 4 个 to 引导的并列的动词短语。

第二步,根据汉语表述习惯和风格,译文采取顺译法。

① 由于电力企业的产品是电能,具有产、供、销同时进行并时刻保持平衡的特点,停止其生产将给国民经济造成巨大损失。

② 电力企业的基本任务就是为用户提供充足、可靠、高质量、廉价的电能,同时,为国家积累大量资金,促进国民经济的发展,改善人民生活。

第三步,回读整个译文,根据上下文做增补和修正。

② 按照汉语习惯使用成对的关联词"由于,因此",句首增译"因此"。

【译文】由于电力企业的产品是电能,具有产、供、销同时进行并时刻保持平衡的特点,停止其生产将给国民经济造成巨大损失,因此电力企业的基本任务就是为用户提供充足、可靠、高质量、廉价的电能,同时,为国家积累大量资金,促进国民经济的发展,改善人民生活。

❖ Although the safety of a nuclear power plant is directly linked to the decision taken at the design stage and to the care taken in pre-construction studies and construction itself, it is only really in operation that nuclear safety can be seen in all its various facets, i.e. the risk of accident due to weaknesses in the design, insufficient quality or operational error.

【分析】第一步,厘清原文的句法结构。

本句为复合句,主要由主句和状语从句组成。其具体句法结构依次是:

① Although the safety of a nuclear power plant is directly linked to the decision <u>taken at the design stage</u> and to the care <u>taken in pre-construction studies and construction itself</u> 为让步状语从句。其中画线部分分别为 decision 和 care 的后置定语。

② it is only really in operation that nuclear safety can be seen in all its various facets, i.e. the risk of accident due to weaknesses in the design, insufficient quality or operational error 为主句。

第二步,根据汉语表述习惯和风格,译文采取顺译法。

① 尽管核电厂的安全与设计阶段的决策、施工前研究和施工本身的谨慎有着直接关系。

② 只有在电厂真正进入运行阶段,才可在各方面看到其核安全状况,如:由于其设计缺陷出现事故危险、质量不佳或操作失误。

第三步,回读整个译文,增补和修正译文。

① 改为"施工前的认真研究以及谨慎施工本身"更为通顺。

② 按照汉语习惯使用成对的关联词"尽管,但",句首增译"但"。

【译文】尽管核电厂安全与设计阶段的决策、施工前的认真研究以及谨慎施工本身有着直接关系,但只有在电厂真正进入运行阶段,才可在各方面看到其核安全状况,如:由于其设计缺陷出现事故危险、质量不佳或操作失误。

2. 倒译法

❖ Various machine parts can be washed very clean and will be as clean as new ones when they are treated by ultrasonics, no matter how dirty and irregularly shaped they maybe.

【分析】第一步,厘清原文的句法结构。

本句为复合句,主要由主句和状语从句组成。其具体句法结构依次是：

① Various machine parts can be washed very clean and will be as clean as new ones 为主句。

② when they are treated by ultrasonics 为时间状语从句。

③ no matter how dirty and irregularly shaped they maybe 为让步状语从句。

第二步,根据汉语表述习惯和风格,译文采取倒译法。

③ 无论多么脏,形状多不规则。

② 当它们被超声波处理后。

① 各种机器零件都可以被清洗得非常干净,甚至像新零件一样。

第三步,回读整个译文,增补和修正译文。

③ 因为译文句序更改,此句句首增加主语"各种机器零件"。

② 改为"当用超声波处理后"更为简洁。

① 此句去掉主语,同时也根据前句改为主动语态更为简洁明了。

【译文】各种机器零件,无论多么脏,形状多不规则,当用超声波处理后,都可以清洗得非常干净,甚至像新零件一样。

❖ Lest one assume that a knowledge of electronic structure enabled chemists to arrange the elecrons in a systematic table, it should be noted that electrons were not known until over 30 years after the table was in use.

【分析】第一步,厘清原文的句法结构。

本句为复合句,主要由主句和状语从句组成。其具体句法结构依次是：

① Lest one assume that a knowledge of electronic structure enabled chemists to arrange the elements in a systematic table,本句为 lest(唯恐、免得、以免)引导的目的状语从句,其中包括画线部分 that 引导的宾语从句。

② it should be noted that electrons were not known until over 30 years after the table was in use.本句为主句,句首 it 是形式主语,其中包括画线部分 that 引导的主语从句,而主语从句又包括 until 引导的时间状语从句。

第二步,根据汉语表述习惯和风格,译文采取倒译法。

② 必须指出,在这个表使用 30 多年以后,人们才知道电子。

① 绝不要认为,电子结构让化学家排成了元素周期表。

第三步,回读整个译文,增补和修正译文。

② 由于调整了原文语序,译文中 table 由"这个表"增补为"元素周期表"。

按照汉语习惯使用成对的关联词"尽管,但",句首增译"但"。

① 为了和前句主动语态一致,此句也改为主动语态"化学家在知道电子结构以后,才能

排成元素周期表。"

【译文】必须指出,在周期表使用 30 多年以后,人们才知道电子。绝不要认为,化学家在知道电子结构以后,才能排成元素周期表。

3. 分译法

❖ With the advent of the space shuttle, it will be possible to put an orbiting solar power plant in stationary orbit 24 000 miles from the earth that would collect solar energy almost continuously and convert this energy either directly to electricity via photovoltaic cells or indirectly with flat plate or focused collectors that would boil a carrying medium to produce steam that would drive a turbine that then in turn would generate electricity.

【分析】第一步,厘清原文的句法结构。

本句的主干 it will be possible 为主系表结构,修饰部分主要是四个定语从句。其具体句法结构依次是:

① With the advent of the space shuttle, it will be possible to put an orbiting solar power plant in stationary orbit 24 000 miles from the earth 为主句。

② that would collect solar energy almost continuously and convert this energy either directly to electricity via photovoltaic cells or indirectly with flat plate or focused collectors 定语从句修饰上文中的 an orbiting solar power plant。

③ that would boil a carrying medium to produce steam 定语从句修饰上文中的 collectors。

④ that would drive a turbine 定语从句指代上文③。

⑤ that then in turn would generate electricity 定语从句修饰上文中的 turbine。

第二步,根据汉语表述习惯和风格,译文采取分译法,也就是将主干部分和四个定语从句分开翻译。

① 随着航天飞机的出现,将有可能在距地球 24 000 英里的静止轨道上放置一个轨道太阳能发电厂。

② 它会几乎持续不断地收集太阳能,并通过光伏电池直接将太阳能转化为电能,或者通过平板或聚焦收集器间接地转化为电能。

③ 这些收集器将煮沸运载介质,产生蒸汽,

④ 驱动涡轮机,

⑤ 然后涡轮机就可以发电了。

第三步,回读整个译文,增补和修正译文。

② "它"改为"该发电厂"意思更加清楚。

③ 改为"这些收集器让运载介质沸腾产生蒸汽"更加通顺。

④ 句首加上"从而"更加连贯。

【译文】随着航天飞机的出现,将有可能在距地球 24 000 英里的静止轨道上放置一个轨道太阳能发电厂。该发电厂可以连续收集太阳能,并通过光伏电池直接将太阳能转化为电能,或者通过平板或聚焦收集器间接地转化为电能。这些收集器让运载介质沸腾产生蒸汽,从而驱动涡轮机,然后涡轮机就可以发电了。

❖ It's easy to think, from the Western perspective, that the great days of engineering were in the past during the era of massive mechanization and urbanization that had its heyday in the nineteenth century and which took the early Industrial Revolution from the eighteenth century right through into the twentieth century which, incidentally, simultaneously improved the health and well-being of the common person with improvements in water supply and sanitation.

【分析】第一步,厘清原文的句法结构。

本句主干为 It's easy to think that..., that 后面的宾语从句中又包含了三个定语从句。其具体句法结构依次是:

① It's easy to think, from the Western perspective, that the great days of engineering were in the past 为主干,其中 it 为形式主语,to think that 为真正的主语,that 后面为宾语从句。

② during the era of massive mechanization and urbanization that had its heyday in the nineteenth century 为时间状语,包括 that 引导宾语从句修饰本句中的 the era。

③ and which took the early Industrial Revolution from the eighteenth century right through into the twentieth century 定语从句修饰上文中的 the era。

④ which, incidentally, simultaneously improved the health and well-being of the common person with improvements in water supply and sanitation 定语从句指代上文③。

第二步,根据汉语表述习惯和风格,译文采取分译法,也就是将主干部分和另外三个定语从句分开翻译。

① 从西方的角度来看,很容易认为人类工程学的辉煌时期是在过去。
② 在十九世纪达到鼎盛的大规模机械化和城市化时代。
③ 使得早期工业革命从十八世纪延续到了二十世纪。
④ 这附带地改善了普通人的健康和福祉,改善了供水和卫生设施。

第三步,回读整个译文,增补和修正译文。

① 联系上下文,翻译成"人类工程学的辉煌时代似乎已经过去了"更为通顺。
② 句首调整为"那时是"更为连贯。
③ 句首加上连接词"也"。
④ 把"附带"改为"间接"更加符合中文表达。把"改善了供水和卫生设施"放在"改善了普通人的健康和福祉"前面,中间加上连词"从而"使得逻辑关系更加清楚。

【译文】从西方的角度来看,很容易认为人类工程学的辉煌时代已经过去了。那时是19世纪达到鼎盛的大规模机械化和城市化时代,也使得早期工业革命从18世纪延续到了20世纪。这间接改善了供水和卫生设施从而改善了普通人的健康和福祉。

4. 综合法

❖ With more money for development of novel designs and public financial support for construction—perhaps as part of a clean energy portfolio standard that lumps in all low-carbon energy sources, not just renewables or a carbon tax—nuclear could be one of the pillars of a three-pronged approach to cutting greenhouse gas emissions: using less energy to do more (or energy efficiency), low-carbon power, and electric cars as long as they are charged with electricity from clean sources, not coal burning.

【分析】第一步,厘清原文的句法结构。

本句的主干为 nuclear could be 主系表结构,修饰部分主要为两个状语和一个插入语。其具体句法结构依次是:

① With more money for development of novel designs and public financial support for construction 为伴随状语。

② perhaps as part of a clean energy portfolio standard that lumps in all low-carbon energy sources, not just renewables or a carbon tax 为插入语,用于解释说明前面的"public financial support for construction"。

③ nuclear could be one of the pillars of a three-pronged approach to cutting greenhouse gas emissions 为主句。

④ using less energy to do more (or energy efficiency), low-carbon power, and electric cars <u>as long as they are charged with electricity from clean sources, not coal burning</u>. 为同位语,用来解释 one of the pillars of a three-pronged approach,其中包括画线部分的条件状语从句。

第二步,根据汉语表述习惯和风格,先顺译,再倒译(原句由于插入语过长,因此放到主句后面翻译),同时也采用分译法。

① 获得更多资金来开发新设计,并获得公共财政的支持用以建设。(名词转译为动词)

③ 核能可以成为三管齐下减少温室气体排放的方法之一。

② 可能作为清洁能源组合标准的一部分,将所有低碳能源纳入其中,而不仅限于可再生能源或碳税制度。

④ 用更少能源做更多的事(提高能源效率),使用低碳能源以及使用电动汽车(前提是它们使用的电能来自清洁能源,而不是煤炭燃烧)。

第三步,回读整个译文,增补潜在信息。

① 句首增加连接词"如果"。

③ 译文不变。

② 分译法句首增加主语"这一方案"。

④ 分译法句首增加主语"这三管齐下的方法包括"。

【译文】如果获得更多资金来开发新设计,并获得公共财政的支持用以建设,核能可以成为三管齐下减少温室气体排放的方法之一。这一方案可能作为清洁能源组合标准的一部分,将所有低碳能源纳入其中,而不仅限于可再生能源或碳税。这三管齐下的方法包括用更少的能源做更多的事(即能源效率提升),使用低碳能源以及使用电动汽车(前提是它们使用的电能来自清洁能源,而不是煤炭燃烧)。

第三节 语篇翻译视角

前面两节具体讨论了能源电力工程项目管理相关的词和句子的翻译技巧,本篇将从语篇视角出发,探讨能源电力工程项目管理的英文文章的特点和翻译技巧。首先,词汇方面,能源电力工程管理涉及能源电力方面的专业术语,要做到语篇中的术语和词语翻译的前后一致性。其次,句子方面,要注意语篇中语义的衔接和连贯。最后,行文方面,要注意语篇中段落之间的逻辑关系。

一、语篇翻译中词语的前后一致性

❖ All power systems have one or more sources of power. For some power systems, the source of power is external to the system but for others it is part of the system itself. Direct current power can be supplied by batteries, fuel cells or photovoltaic cells. Alternating current power is typically supplied by a rotor that spins in a magnetic field in a device known as a turbo generator in a power station. There have been a wide range of techniques used to spin a turbine's rotor, from superheated steam heated using fossil fuel (including coal, gas and oil) to water itself (hydroelectric power) and wind (wind power). Even nuclear power typically depends on water heated to steam using a nuclear reaction.

【译文】所有的电力系统都有一个或多个电源。对某些电力系统来说,电源是来自系统的外部,但对另外一些系统来说,电源则是系统本身的一部分。直流电可以通过电瓶、燃料电池或光伏电池供电。交流电主要是通过发电厂的汽轮发电机转子在磁场中旋转而生产电。汽轮机转子旋转所使用的技术范围很广,包括使用化石燃料(如煤、石油、天然气)加热而产生的过热蒸汽、水(水电)和风(风电)。甚至核电也主要依赖水,利用核反应将水加热成蒸汽。

【注解】power 一词有多种含义,可以表示"功率;能;力;电;电力"等。本文中 power 出现了 10 次,在不同语境中有不同的意义,翻译中需要准确定义并确保前后一致。本篇中 power systems 出现 2 次,均译为"电力系统";source of power 出现 2 次,均译为"电源";direct current power 和 alternating current power 各出现 1 次,分别译为"直流电"和"交流电";hydroelectric power、wind power 和 nuclear power 中各出现 1 次,分别译为水电、风电和核电;power station 出现 1 次,译为"发电厂"。

二、语篇翻译中句子的衔接和连贯

❖ With few exceptions, however, this is not the case with independent voltage and current sources. Although an actual battery can often be thought of as an ideal voltage source, other non-ideal independent sources are approximated by a combination of circuit elements. Among these elements are the dependent sources, which are not discrete components as are many resistors and batteries but are in a sense part of electronic devices like transistors and operational amplifiers. But don't try to peel open a transistor's metal so that you can see a little diamond-shaped object. The dependent source is a theoretical element that is used to help describe or model the behavior of various electrical devices.

【译文】然而,除了极少数几个例子外,对于独立的电压源和电流源来说,情况就并不是这样。虽然一个实际的电池往往可以看成是一个理想电压源，但是其他的非理想独立源可以用一组电路元件加以近似。这些元件中有非独立源，它们并不像许多电阻器和电池那样属于分立部件，而在某种意义上属于像晶体管和运算放大器这样的电子器件的一部分。不过不要剥开晶体管壳，里面有一个小的菱形状物体。非独立源是用来帮助描述各种电器件的性能或对其进行建模的一种理论元件。

【注解】英文语篇的特点体现在形式上的衔接和语义上的连贯。能源电力工程项目管理类的语篇由于以客观陈述为主，通常语篇组织严谨，结构紧凑。句子之间通常用过渡词衔接，比如原文中的however, although, but表达了转折的关系，在翻译中一般直译。语义上的连贯除了通过准确翻译过渡词，还要注意代词的翻译以及中英文的差别。比如原文第三句中的which在译文中翻成"它们"体现语义的连贯，原文中的you can see在译文中简化为"里面有"，使得句子表达更加紧凑连贯。

三、语篇翻译的逻辑性

- While green energy's footprint continues to increase as the fastest growing segment within the global energy mix, it still trails significantly behind conventional, high-carbon energy options due to efficiency and capacity hurdles. These limitations are preventing green energy from becoming a predominant, mainstream energy option. What other scalable, non-CO2 emitting form of energy could help us close the gap until green energy becomes a large-scale reality for us? Could nuclear energy's carbon-free profile, proven efficiency and scalability make nuclear power a transition candidate, and possibly another viable, widely accepted energy option for the future?

- In addition to their zero-emission blueprint, the approximately 450 nuclear power plants operate today at full capacity greater than 90% of the time compared to 50% for coal and 25% for solar plants. However, only 10% of the total electricity demand worldwide is supplied by nuclear power plants. Why hasn't nuclear energy grown faster over the years?

- Although a proven and economical option for energy production, nuclear energy carries a controversial image due to the risks associated with radioactivity and its impact on the environment. The Chernobyl and Fukushima events reminded us that atomic fission requires flawless control and vigilance and that small incidents can turn into major catastrophes.

【译文】绿色能源作为全球能源结构中增长最快的部分，虽然仍然在持续发展，但由于效率和产能障碍，它仍然明显落后于传统的高碳能源选项。这些限制阻碍了绿色能源成为主导的主流能源选择。绿色能源在现实中大规模应用之前，还有什么其他可扩展的、不排放二氧化碳的能源可以帮助我们缩小这一差距吗？核能的无碳特性、已证明的效率和可扩展性能否使其成为过渡的候选能源，并可能成为另一个未来可行的并被广泛接受的能源选项吗？

除了零排放的蓝图，目前约450座核电站在运行，其满负荷运行时间超过90%，而燃煤电厂和太阳能电厂的满负荷运行时间分别为50%和25%。然而,全世界只有10%的电力需求是由核电站提供的。为什么核能多年来没有更快地增长?

尽管核能是一种经过验证的、经济的能源,但由于其与放射性有关的风险及对环境的影响,核能的形象备受争议。切尔诺贝利和福岛事件提醒我们,原子裂变需要完美的控制和警惕,否则小事件可能变成重大灾难。

【注解】语篇中三个段落之间的逻辑关系是依次紧密联系的,通过提出问题和回答问题的方式,讨论了绿色能源和核能作为替代能源的可行性和限制。同时,段落间的衔接词也体现了三个段落之间多次转折关系。比如第一个段落由绿色能源效率和产能的限制,引出核能是否可以作为能源选择的问题。第二个段落同时也引出了核能增长缓慢的问题。第三个段落则针对前两段的问题进行解答,指出了核能面临的争议和风险。在语篇翻译中,译者需要从形式和内容上厘清句子、段落间的逻辑关系,从而构建文理通顺的语篇。

- The water resources are always protected throughout the utilization of hydropower energy by human beings. From the diversion irrigating project of Dujiangyan in China to the urban water-supply system in ancient Rome, the effective flood control by building dams and water resource utilization by constructing diversion channels have become the important component of the human being's civilization history of thousands of years. In 1878 of the Industrial Revolution period in France, the first hydropower station in the world was built, which had a significance in the historical process of water resource development and utilization. Since then, with the advantages of clean, effective, reusable and low-cost characteristics, hydropower energy has become the engine for industrialization and will develop and flourish with the progress of modern civilization. Since the twentieth century, the construction of high dam and large-sized hydropower projects has become the sign of economic development and social advancement, as well as the embodiment of capability for the international community to advance the modernization process and for human beings to protect and unitize natural resources. Undoubtedly, hydropower energy has become the important component of present civilization of human society, which is also the key approach to solve the problem of unbalance of water resources.

- In recent years, with the acceleration of worldwide industrialization process, the series of ecological problems of global warming caused by the large consumption of fossil energy, greenhouse gas emission such as carbon dioxide have set off an alarm to the human beings. It has become a consensus and research orientation in the international society to vigorously promote "low-carbon energy", develop "low-carbon economy", and advocate "low-carbon living", in which the hydropower energy has received active promotion and development priority as a renewable energy in most of the countries owing to its mature technologies, developable features and rich reserves.

【译文】人类利用水力发电一直伴随着对水资源的保护利用。从中国的都江堰引水灌溉到古罗马的城市供水系统，通过建筑堤坝有效地控制洪水和修渠引水利用水资源已经成为人类几千年文明史的重要组成部分。工业革命时期的1878年法国建成了世界上第一座水电站，这在水能资源开发利用进程中具有划时代的意义。此后水电以其清洁、高效、可重复利用和低运行成本等众多优势，成为工业化的推动力，并伴随着现代文明的进步而蓬勃发展。20世纪以来，建造高坝大库和大型水电工程成为经济发展和社会进步的标识，成为国际社会推进现代化进程和展现人类保护利用自然资源能力的体现。毋庸置疑，水力发电已经成为当今人类社会文明的重要组成部分，是解决水资源丰枯不均矛盾的重要途径。

近年来，随着世界范围内工业化进程的加快，由化石能源被大量消耗、二氧化碳等温室气体排放、全球气候变暖引发的一系列生态问题向世人敲响着警钟。大力推广"低碳能源"、大力发展"低碳经济"、大力倡导"低碳生活"，已成为国际社会的普遍共识和求索方向，其中水电作为目前技术最成熟、最具开发性和资源最丰富的可再生能源，得到了绝大多数国家的积极提倡和优先发展。

【注解】语篇中两个段落之间的逻辑关系主要是通过内容体现的。第一个段落按照时间顺序介绍了水力发电是人类文明史的重要组成部分，并指出水力能源的优势和发展趋势。而第二个段落则提及近年来全球工业化进程加快所引发的生态问题，从而强调水力发电作为可再生能源在解决这些问题中扮演着重要角色。在能源电力工程项目管理的翻译中，译者要先从语篇角度深入理解原文的逻辑行文，然后再按照汉语的习惯组织语言，才可能形成结构严谨、语句达意的译文。

第五章 能源电力工程管理英语翻译实践

本章提供了能源电力工程相关英语语篇,以帮助读者利用前文介绍的翻译技巧进行有效的翻译实践。本章分为两个小节:第一节的语篇主要涉及能源电力工程管理基本知识;第二节的语篇主要涉及新时代中国能源电力的发展情况。

第一节 能源电力工程管理基本知识

本节有9篇文章,以英中双语形式展现,在每个课文后面有注释和参考译文,供学习者研读课文使用。课后设翻译练习题,供学生翻译练习使用。

Text 1 Renewable Energy—Powering a Safer Future

A large chunk of the greenhouse gases that blanket the Earth and trap the sun's heat are generated through energy production, by burning fossil fuels to generate electricity and heat. Fossil fuels, such as coal, oil and gas, are by far the largest contributor to global climate change, accounting for over 75 percent of global greenhouse gas emissions and nearly 90 percent of all carbon dioxide emissions.

The science is clear: to avoid the worst impacts of climate change, emissions need to be reduced by almost half by 2030 and reach net-zero by 2050. To achieve this, we need to end our reliance on fossil fuels and invest in alternative sources of energy that are clean, accessible, affordable, sustainable, and reliable. Renewable energy sources—which are available in abundance all around us, provided by the sun, wind, water, waste, and heat from the Earth-are replenished by nature and emit little to no greenhouse gases or pollutants into the air. Fossil fuels still account for more than 80 percent of global energy production, but cleaner sources of energy are gaining ground. About 29 percent of electricity currently comes from renewable sources.

Here are five reasons why accelerating the transition to clean energy is the pathway to a healthy, livable planet today and for generations to come.

1. Renewable energy sources are all around us

About 80 percent of the global population lives in countries that are net-importers of fossil fuels—that's about 6 billion people who are dependent on fossil fuels from other countries, which makes them vulnerable to geopolitical shocks and crises.

In contrast, renewable energy sources are available in all countries, and their potential is yet to be fully harnessed. The International Renewable Energy Agency (IRENA) estimates that 90 percent of the world's electricity can and should come from renewable energy by 2050.

Renewables offer a way out of import dependency, allowing countries to diversify their economies and protect them from the unpredictable price swings of fossil fuels, while driving inclusive economic growth, new jobs, and poverty alleviation.

2. Renewable energy is cheaper

Renewable energy actually is the cheapest power option in most parts of the world today. Prices for renewable energy technologies are dropping rapidly. The cost of electricity from solar power fell by 85 percent between 2010 and 2020. Costs of onshore and offshore wind energy fell by 56 percent and 48 percent respectively.

Falling prices make renewable energy more attractive all around—including to low- and middle-income countries, where most of the additional demand for new electricity will come from. With falling costs, there is a real opportunity for much of the new power supply over the coming years to be provided by low-carbon sources.

Cheap electricity from renewable sources could provide 65 percent of the world's total electricity supply by 2030. It could decarbonize 90 percent of the power sector by 2050, massively cutting carbon emissions and helping to mitigate climate change.

Although solar and wind power costs are expected to remain higher in 2022 and 2023 than pre-pandemic levels due to general elevated commodity and freight prices, their competitiveness actually improvesdue to much sharper increases in gas and coal prices, says the International Energy Agency (IEA).

3. Renewable energy is healthier

According to the World Health Organization (WHO), about 99 percent of people in the world breathe air that exceeds air quality limits and threatens their health, and more than 13 million deaths around the world each year are due to avoidable environmental causes, including air pollution.

The unhealthy levels of fine particulate matter and nitrogen dioxide originate mainly from the burning of fossil fuels. In 2018, air pollution from fossil fuels caused $2.9 trillion in health and economic costs, about $8 billion a day.

Switching to clean sources of energy, such as wind and solar, thus helps address not only climate change but also air pollution and health.

4. Renewable energy creates jobs

Every dollar of investment in renewables creates three times more jobs than in the fossil fuel industry. The IEA estimates that the transition towards net-zero emissions will lead to an overall increase in energy sector jobs: while about 5 million jobs in fossil fuel production could be lost by 2030, an estimated 14 million new jobs would be created in clean energy, resulting in a net gain of 9 million jobs.

In addition, energy-related industries would require a further 16 million workers, for instance to take on new roles in manufacturing of electric vehicles and hyper-efficient appliances or in innovative technologies such as hydrogen. This means that a total of more than 30 million jobs could be created in clean energy, efficiency, and low-emissions technologies by 2030.

Ensuring a just transition, placing the needs and rights of people at the heart of the energy transition, will be paramount to make sure no one is left behind.

5. Renewable energy makes economic sense

About $5.9 trillion was spent on subsidizing the fossil fuel industry in 2020, including through explicit subsidies, tax breaks, and health and environmental damages that were not priced into the cost of fossil fuels.

In comparison, about $4 trillion a year needs to be invested in renewable energy until 2030—including investments in technology and infrastructure—to allow us to reach net-zero emissions by 2050.

The upfront cost can be daunting for many countries with limited resources, and many will need financial and technical support to make the transition. But investments in renewable energy will pay off. The reduction of pollution and climate impacts alone could save the world up to $4.2 trillion per year by 2030.

Moreover, efficient, reliable renewable technologies can create a system less prone to market shocks and improve resilience and energy security by diversifying power supply options.

"The good news is that the lifeline is right in front of us," says UN Secretary-General António Guterres, stressing that renewable energy technologies like wind and solar already exist today, and in most cases, are cheaper than coal and other fossil fuels. We now need to put them to work, urgently, at scale and speed.

The Secretary-General outlines five critical actions the world needs to prioritize now to transform our energy systems and speed up the shift to renewable energy—"because without renewables, there can be no future."

1. Make renewable energy technology a global public good

For renewable energy technology to be a global public good—meaning available to all, and not just to the wealthy—it will be essential to remove roadblocks to knowledge sharing and technological transfer, including intellectual property rights barriers.

Essential technologies such as battery storage systems allow energy from renewables, like solar and wind, to be stored and released when people, communities and businesses need power. They help to increase energy system flexibility due to their unique capability to quickly absorb, hold and re-inject electricity, says the International Renewable Energy Agency.

Moreover, when paired with renewable generators, battery storage technologies can provide reliable and cheaper electricity in isolated grids and to off-grid communities in remote locations.

2. Improve global access to components and raw materials

A robust supply of renewable energy components and raw materials is essential. More widespread access to all the key components and materials—from the minerals needed to produce wind turbines and electricity networks, to electric vehicles—will be key.

It will take significant international coordination to expand and diversify manufacturing capacity globally. Moreover, greater investments are needed to ensure a just transition—including in people's skills training, research and innovation, and incentives to build supply chains through sustainable practices that protect ecosystems and cultures.

3. Level the playing field for renewable energy technologies

While global cooperation and coordination is critical, domestic policy frameworks must urgently be reformed to streamline and fast-track renewable energy projects and catalyze private sector investments.

Technology, capacity and funds for renewable energy transition exist, but there needs to be policies and processes in place to reduce market risk and enable and incentivize investments—including through streamlining the planning, permitting and regulatory processes, and preventing bottlenecks and red tape. This could include allocating space to enable large-scale build-outs in special Renewable Energy Zones.

Nationally Determined Contributions, countries' individual climate action plans to cut emissions and adapt to climate impacts, must set 1.5C aligned renewable energy targets—and the share of renewables in global electricity generation must increase from today's 29 percent to 60 percent by 2030.

Clear and robust policies, transparent processes, public support and the availability of modern energy transmission systems are key to accelerating the uptake of wind and solar energy technologies.

4. Shift energy subsidies from fossil fuels to renewable energy

Fossil-fuel subsidies are one of the biggest financial barriers hampering the world's shift to renewable energy. The International Monetary Fund (IMF) says that about $5.9 trillion was spent on subsidizing the fossil fuel industry in 2020 alone, including through explicit subsidies, tax breaks, and health and environmental damages that were not priced into the cost of fossil fuels. That's roughly $11 billion a day.

Fossil fuel subsidies are both inefficient and inequitable. Across developing countries, about half of the public resources spent to support fossil fuel consumption benefits the richest 20 percent of the population, according to the IMF.

Shifting subsidies from fossil fuels to renewable energy not only cuts emissions, it also contributes to the sustainable economic growth, job creation, better public health and more equality, particularly for the poor and most vulnerable communities around the world.

5. Triple investments in renewables

At least $4 trillion a year needs to be invested in renewable energy until 2030—including investments in technology and infrastructure—to allow us to reach net-zero emissions by 2050.

Not nearly as high as yearly fossil fuel subsidies, this investment will pay off. The reduction of pollution and climate impact alone could save the world up to $4.2 trillion per year by 2030.

The funding is there—what is needed is commitment and accountability, particularly from the global financial systems, including multilateral development banks and other public and private financial institutions, that must align their lending portfolios towards accelerating the renewable energy transition.

In the Secretary-General's words, "Renewables are the only path to real energy security, stable power prices and sustainable employment opportunities."

Notes

（1）greenhouse gas 温室气体：指能够在大气中吸收和重新辐射地球表面的红外线辐射气体，这些气体包括二氧化碳、甲烷、氟氯碳化物等。

（2）fossil fuels 化石燃料：指煤炭、石油和天然气等能源，它们形成于数百万年的生物化学过程中。

（3）net-zero 零增长：指排放的温室气体总量和从环境中移除的温室气体总量持平。

（4）inclusive economic growth 包容性经济增长：指经济增长的过程中各种社会群体都能够分享经济增长和从中受益的情况。

（5）just transition 公正过渡，公正转型：指实现向可持续能源转变的过程中，保障劳工和社会的公平和合理的方式。它是一个框架，旨在实现向低碳经济的公平和可持续转型，由工会提出，并得到环保非政府组织的支持。通过将其纳入2009年联合国气候变化会议的谈判文本，它得到了更广泛的关注。

（6）streamline 简化：指通过精简程序和流程，使其更加高效和简便。

（7）red tape 繁文缛节，官僚作风：源自用红色或粉红色的带子捆扎公文的习俗。

（8）multilateral development banks 多边发展银行：指由多个国家共同设立的开发银行，致力于促进发展和减轻贫困。

参考译文

第1课　可再生能源：为更安全的未来提供动力

覆盖地球并吸收太阳热量的温室气体中，有很大一部分是通过能源生产产生的，即通过燃烧化石燃料来发电和供热产生的。化石燃料，如煤炭、石油和天然气，是迄今为止全球气候变化最大的促成因素，占全球温室气体排放量的 75% 以上，占所有二氧化碳排放量的近 90%。

科学是明确的：为了避免气候变化的最坏影响，需要到 2030 年将排放量减少近一半，到 2050 年达到净零排放。为了实现这一目标，我们需要结束对化石燃料的依赖，并投资于清洁、可获得、负担得起、可持续和可靠的替代能源。我们周围到处都有大量可再生能源，由太阳、风、水、废物和地球的热量提供，由大自然补充，几乎不向空气中排放温室气体或污染物。化石燃料仍占全球能源生产的 80% 以上，但清洁能源正在普及。目前约 29% 的电力来自可再生能源。

加速向清洁能源过渡是当今和未来几代人建设健康、宜居地球的唯一途径，其原因有如下五个。

1. 可再生能源无处不在

全球大约 80% 的人口生活在化石燃料的净进口国，大约 60 亿人依赖于来自其他国家的化石燃料，这使得他们容易受到地缘政治冲击和危机的影响。

相比之下，所有国家都有可再生能源，但其潜力尚未得到充分利用。国际可再生能源署 (IRENA) 估计，到 2050 年，世界上 90% 的电力可以而且应该来自可再生能源。

可再生能源提供了一条摆脱进口依赖的途径，使各国能够实现经济多样化，保护它们免受化石燃料不可预测的价格波动的影响，同时推动包容性经济增长、新增就业和减贫。

2. 可再生能源更便宜

可再生能源实际上是当今世界大多数地区最便宜的能源选择。可再生能源技术的价格正在迅速下降。2010 年至 2020 年间，太阳能发电的成本下降了 85%。陆上和海上风能的成本分别下降了 56% 和 48%。

价格下跌使可再生能源在各个方面都更具吸引力，包括对中低收入国家而言也是如此，全球对新电力的额外需求将主要来自这些国家。随着成本的下降，未来几年大部分新的电力供应都有真正的机会由低碳能源提供。

到 2030 年，来自可再生能源的廉价电力将提供世界总电力供应的 65%。到 2050 年，它可以使 90% 的电力部门脱碳，大规模减少碳排放，并有助于减缓气候变化。

国际能源署 (IEA) 表示，尽管由于大宗商品和货运价格普遍上涨，预计 2022 年和 2023 年太阳能和风能成本仍将高于疫情前的水平，但由于天然气和煤炭价格大幅上涨，它们的竞争力实际上有所提高。

3. 可再生能源更健康

据世界卫生组织 (WHO) 统计，世界上约 99% 的人呼吸的空气超过空气质量限值，威胁到他们的健康，全球每年有超过 1300 万人的死亡是由于包括空气污染在内的可避免的环境原因导致的。

有害健康的微粒物质和二氧化氮主要来自矿物燃料的燃烧。2018 年，化石燃料造成的空气污染造成了 2.9 万亿美元的健康和经济成本，约为每天 80 亿美元。

因此，改用清洁能源，如风能和太阳能，不仅有助于应对气候变化，而且有助于解决空气污染和健康问题。

4. 可再生能源创造就业机会

对可再生能源每一美元投资所创造的就业机会,是化石燃料行业的三倍。国际能源署估计,向净零排放的过渡将导致能源部门就业机会的总体增加:虽然到2030年矿物燃料生产方面可能失去约500万个工作岗位,但估计清洁能源方面将创造1400万个新的工作岗位,从而净增加900万个工作岗位。

此外,与能源有关的行业还需要1600万工人,例如,在电动汽车和超高效电器的制造或氢等创新技术方面承担新的角色。这意味着到2030年,清洁能源、增效和低排放技术领域总共可以创造3000多万个就业岗位。

确保一个公正的过渡,将人民的需求和权利置于能源转型的核心位置,对于确保不让任何一个人掉队至关重要。

5. 可再生能源具有经济意义

2020年,大约5.9万亿美元用于补贴化石燃料行业,包括通过明确的补贴、税收减免以及未计入化石燃料成本的健康和环境损害。

相比之下,到2030年,每年需要约4万亿美元投资于可再生能源,包括技术和基础设施投资,通过这种方式才能让我们到2050年实现净零排放。

对于许多资源有限的国家来说,前期成本可能令人望而生畏,许多国家将需要财政和技术支持来实现过渡。但对可再生能源的投资会有回报。到2030年,仅减少污染和气候影响一项就可以为世界每年节省高达4.2万亿美元。

此外,高效、可靠的可再生能源技术可以创造一个不易受市场冲击影响的系统,并通过使电力供应选择多样化来提高韧性和能源安全。

联合国秘书长安东尼奥·古特雷斯(António Guterres)表示:"好消息是,救生索就在我们面前。"他强调,如今已具备风能、太阳能等可再生能源技术,而且在大多数情况下,这些可再生能源比煤炭等化石燃料更便宜。我们现在迫切地需要让它们大规模地、快速地发挥作用。

秘书长提炼了世界目前需要优先采取的五项关键行动,以改造我们的能源系统,加快向可再生能源的转型,"因为没有可再生能源,我们就没有未来"。

1. 使可再生能源技术成为全球公益品

可再生能源技术成为全球公益品就意味着所有人都可获得,而不仅仅在富人手中,为此就必须扫除知识共享和技术转让方面的障碍,包括知识产权壁垒。

电池储存系统等必要技术能够储存太阳能、风能等可再生能源,并在人们、社区和企业需要电力时释放出来。国际可再生能源署表示,由于这些技术具有快速吸收、保存和重新注入电力的独特能力,它们有助于提高能源系统的灵活性。

此外,当与可再生发电机配合使用时,电池储能技术可以为孤立电网和位于偏远地区的离网社区提供可靠且廉价的电力。

2. 改进全球获取组件和原材料的渠道

供应充足的可再生能源组件和原材料是至关重要的。重点在于所有关键组件和材料(从生产风轮机和电网所需的矿物到电动汽车)更广泛的普及性。

要在全球范围内扩增制造力并使之多样化,需要进行大量的国际协调。此外,需要加大投资以确保公正的过渡,包括在人们的技能培训、研究和创新方面,以及通过保护生态系统和文化的可持续做法来建立供应链的激励措施。

3. 为可再生能源技术创造平等竞争环境

虽然全球合作与协调至关重要,但国内政策框架也必须尽快改革,精简并加快推动可再生能源项目,促进私营部门投资。

虽然具备了可再生能源转型的技术、产能和资金,但仍需要有政策和程序到位来降低市场风险,促进和激励投资,包括通过简化规划、许可和监管程序,防止阻塞现象和繁文缛节。这可能包括分配空间以便在特殊的可再生能源区进行大规模扩建。

国家自主贡献是各国为减少排放和适应气候影响而制定的单独气候行动计划,必须设定与1.5摄氏度目标相一致的可再生能源具体目标。到2030年,全球范围内可再生能源发电的比例必须从目前的29%增加到60%。

明确而有力的政策、透明的程序、公众支持和可用的现代能源传输系统是加快采用风能和太阳能技术的关键所在。

4. 将能源补贴从化石燃料转为可再生能源

化石燃料补贴是世界可再生能源转型中首要的金融障碍之一。国际货币基金组织(IMF)表示,仅2020年就有约5.9万亿美元用于补贴化石燃料行业,包括通过明确的补贴、税收减免以及未计入化石燃料成本的健康和环境损害,每天大约是110亿美元。

化石燃料补贴既不高效也不公平。根据国际货币基金组织的数据,在发展中国家,用于支持化石燃料消费的公共资源中,约有一半使最富有的20%人口获益。

将补贴从化石燃料转向可再生能源,不仅能减少排放,还有助于经济可持续增长、创造就业、改善公共卫生和促进平等,特别是对世界各地的贫困和最脆弱的社区而言尤为助益。

5. 可再生能源的投资增至三倍

到2030年,每年的可再生能源投资至少需要4万亿美元(包括技术和基础设施投资),才能让我们到2050年实现净零排放。

尽管远非每年的化石燃料补贴一样花费高昂,但这项投资将获得回报。到2030年,每年仅减少污染和气候的影响方面就可以为全世界节省多达4.2万亿美元。

资金已经到位,还需要承诺和问责制度,特别是来自全球金融体系,包括多边开发银行和其他公共或私营金融机构,它们必须调整其贷款组合,以加快可再生能源转型。

按秘书长所言:"可再生能源是真正实现能源安全、稳定电力价格和可持续就业机会的唯一途径。"

英翻中练习

(1) You flip a switch. Coal burns in a furnace, which turns water into steam. That steam spins a turbine, which activates a generator, which pushes electrons through the wire. This current propagates through hundreds of miles of electric cables and arrives at your home.

All around the world, countless people are doing this every second—flipping a switch, plugging in, pressing an "on" button. So how much electricity does humanity need? The amount we collectively use is changing fast, so to answer this question, we need to know not just how much the world uses today, but how much we'll use in the future.

The first step is understanding how we measure electricity. It's a little bit tricky. A joule is a unit of energy, but we usually don't measure electricity in just joules. Instead, we measure it in watts. Watts tell us how much energy, per second, it takes to power something. One joule per second equals one watt. It takes about 0.1 watt to power a smart phone, a thousand to power your house, a million for a small town, and a billion for a mid-size city.

As of 2020, it takes 3 trillion watts to power the entire world. But almost a billion people don't have access to reliable electricity. As countries become more industrialized and more people join the grid, electricity demand is expected to increase about 80% by 2050.

That number isn't the complete picture. We'll also have to use electricity in completely new ways. Right now, we power a lot of things by burning fossil fuels, emitting an unsustainable amount of greenhouse gases that contribute to global warming. We'll have to eliminate these emissions entirely to ensure a sustainable future for humanity. The first step to doing so, for many industries, is to switch from fossil fuels to electric power. We'll need to electrify cars, switch buildings heated by natural gas furnaces to electric heat pumps, and electrify the huge amount of heat used in industrial processes. So all told, global electricity needs could triple by 2050.

(2) We'll also need all that electricity to come from clean energy sources if it's going to solve the problems caused by fossil fuels. Today, only one third of the electricity we generate comes from clean sources. Fossil fuels are cheap and convenient, easy to ship, and easy to turn into electricity on demand. So how can we close the gap?

Wind and solar power work great for places with lots of wind and sunshine, but we can't store and ship sunlight or wind the way we can transport oil. To make full use of energy from these sources at other times or in other places, we'd have to store it in batteries and improve our power grid infrastructure to transport it long distances.

Meanwhile, nuclear power plants use nuclear fission to generate carbon-free electricity. Though still more expensive than plants that burn fossil fuels, they can be built anywhere and don't depend on intermittent energy sources like the sun or wind. Researchers are currently working to improve nuclear waste disposal and the safety of nuclear plants.

There's another possibility we've been trying to crack since the 1940s: nuclear fusion. It involves smashing light atoms together, so they fuse, and harnessing the energy this releases. Accidents aren't a concern with nuclear fusion, and it doesn't produce the long-lived radioactive waste fission does. It also doesn't have the transport concerns associated with wind, solar, and other renewable energy sources. A major breakthrough here could revolutionize clean energy.

The same is true of nuclear fission, solar, and wind. Breakthroughs in any of these technologies, and especially in all of them together, can change the world: not only helping us triple our electricity supply, but enabling us to sustain it.

Text 2　The Working Principle of a Thermal Power Plant

Thermal power plants help meet almost half of the world's power demand. They use water as a working fluid. Today's thermal power plants are capable of running under great efficiency by conforming to stringent environmental standards. We will see how a coal-based thermal power plant achieves this in a detailed step by step manner.

By turning the shaft of a generator, electricity will be generated. The generator derives motion from a steam turbine, the heart of the power plant. In order to turn the steam turbine, you have to supply a high pressure and high temperature steam at the inlet of the turbine. As the turbine absorbs energy from the high energy fluid, its pressure and temperature drop toward the outlet, and you can take a closer look at the uniquely shaped steam turbine rotor blades. High capacity power plants often use different stages of steam turbines such as high pressure turbine, intermediate pressure turbine and low pressure turbines. So now we have met our objective and have produced electricity from the generator.

If we can bring the low pressure and low temperature steam back to their original state which were of a much higher pressure and temperature, we can repeat the process. The first step is to raise the pressure. We can use a compressor for this purpose. But compressing steam is a highly energy intensive process. And such a power plant will not be efficient at all. The easy way is to convert the steam into liquid and boost the pressure. For this purpose, we introduce condenser heat exchangers, which sit beneath the low pressure turbine. In the condenser, a stream of cold water flows through the tubes. The steam rejects heat to this liquid stream and becomes condensed. Now we can use a pump to increase the pressure of this feed water. Typically multi-stage centrifugal pumping is used for this purpose. That way the pressure will revert to its original state. The next task is to bring the temperature back to its original value. For this purpose, heat is added to the exit of the pump with the help of a boiler.

High capacity power plants generally use a type of boiler called a water tube boiler. Pulverized coal is then burned inside the boiler. The incoming water initially passes through any economizer session. Here, the water will capture energy from the flue gas. The water flows through a down-comer, and then through water walls, where it transforms into steam. The pure steam is separated at a steam drop. Now the working fluid is back to its original state—high pressure and high temperature. This steam can be fed back into the steam turbine and the cycle can be repeated over and over again for continuous power production.

But a power plant working on this basic RankineCycle will have a very low efficiency and a low capacity. We can increase the performance of the power plant considerably with the help of a few simple techniques. In case of super heating, even after the liquid has been converted into steam, even more heat is added and with that the steam becomes super heated. The higher the temperature of the steam, the more efficient the cycle. Just remember the Carnot's Theorem of maximal thermal efficiency is possible. But the steam turbine material will not withstand temperatures of more than 600°C, so super heating is limited to that threshold. The temperature of the steam decreases as it flows along the rows of the blade. Consequently, a great way to increase the efficiency of the power plant is to add more heat after the first turbine stage. This is known as reheating and it will increase the temperature of the steam again, leading to a higher power output and greater efficiency. The low pressure sides of the power plant are prone to sucking the atmospheric air, even with sophisticated ceiling arrangements. The dissolved gases in the feed water will spoil the boiler material overtime. To remove these dissolved gases, an open feed water heater is introduced. Hot steam from the turbine is mixed into the feed water. Steam bubbles generated will absorb the dissolved gases. The mixing also pre-heats the feed water, which helps improve the efficiency of the power plant to an even greater extent. All these techniques make the modern power plant work under an efficiency rate of 40 – 45%.

Now we'll take a look at how heat addition and heat rejection are executed in an actual power plant. The cold liquid is supplied at the condenser with the help of a cooling tower. The heated-up water from the condenser outlet is sprayed in the cooling tower which induces a natural air draft and the sprayed water loses heat. This is how a colder liquid is always provided at the condenser inlet. At the heat addition side the burning coal produces many pollutants. We cannot release these pollutants directly into the atmosphere. So before transferring them to a stack, the exhaust gases are cleared in an electrical static precipitator. The ESP uses plates with high bullet static electricity to absorb the pollutant particles.

Notes

(1) thermal power plants 火力发电厂、热力发电厂:利用燃烧燃料产生热能,并将其转化为电能的发电设施。

(2) working fluid 工作流体:指在动力系统中用于传输能量和进行功的流体,如水、水蒸气、蒸气等。

(3) shaft 轴:在机械设备中用以传输动力或承受转动力的旋转部件。

(4) steam turbine 汽轮机:利用高温高压蒸气的喷射力驱动汽轮机旋转,并产生动力的发电设备中的核心部件。

(5) rotor blade 转子叶片:指汽轮机转子上的刀片状部件,用于将流体的能量转化为旋转能量。

(6) high pressure turbine 高压汽轮机:汽轮机中的一种,负责处理高压蒸气的汽轮机。

(7) condenser heat exchanger 冷凝器热交换器：用于将蒸气冷却并转化成液体的换热装置。

(8) feed water 进给水：进入锅炉的水，用于产生蒸气。

(9) centrifugal pump 离心泵：一种通过离心力将液体输送到高压区域的泵。

(10) water tube boiler 水管锅炉：一种锅炉，其中水在管壁内流动，被燃烧的煤炭加热产生蒸气。

(11) pulverized coal 煤粉

(12) economizer session 节能站

(13) flue gas 烟气：燃烧后产生的气体，在锅炉中被用来加热水。

(14) down-comer 下导管：指在锅炉中将水从上部输送到下部的管道。

(15) water wall 水壁、水冷壁：锅炉内部的冷却结构，通过其中的水来吸收热量。

(16) steam drop 蒸气分离器：用于将混合物中的水与蒸气分离的设备。

(17) superheating 过热：在液体已经转化为蒸气后，继续向其中加热，使蒸气的温度升高。

(18) Carnot's Theorem 卡诺定理：热力学中关于理想循环最大热效率的定理。

(19) threshold 阈值：指某一特定条件或限制的最小或最大值。

(20) ceiling arrangement 天花板设计：设备中采用的高级空气密封或防漏装置。

(21) open feed water heater 开式给水加热器：用于为锅炉提供预热水的设备。

(22) heat rejection 散热：将系统中的热能排出的过程，通常是通过与冷却介质接触，使热能传递到冷却介质并被带走。

(23) cooling tower 冷却塔：用于将热水或其他工质与大气进行热交换的设备，以降低工质的温度。

(24) electrical static precipitator 电除尘器：利用静电原理去除烟气中颗粒污染物的设备。

(25) plates with high bullet static electricity 高电压静电板：具有高电压静电的板状部件，用于吸附烟气中的污染物颗粒。

参考译文

第 2 课　火电厂工作原理

热电厂帮助满足全球近一半的电力需求。它们使用水作为工作流体。如今的热电厂通过符合严格的环境标准，能够以高效率运行。我们将详细介绍燃煤热电厂是如何逐步实现这一目标的。

通过转动一台发电机的轴，我们就能够发电。发电机从汽轮机中获取动能，汽轮机是热电厂的核心。为了使汽轮机转动，需要在涡轮进气口处提供高压和高温的蒸气。当涡轮从高能流体中吸收能量时，在出口处其压力和温度会降低，您可以更仔细地观察到形状独特的汽轮机转子叶片。高装机容量的发电厂通常使用不同级别的汽轮机，如高压汽轮机、中压汽轮机和低压汽轮机。现在，我们已经达到了我们的目标。我们通过发电机产生了电力。

如果我们能够将低压、低温的蒸气恢复到其原始状态，即更高的压力和温度，我们就可以重复这个过程。第一步是提高压力。我们可以使用压缩机来实现这一目的。但是,压缩蒸气是一个高能耗的过程。这样的电厂效率将非常低下。更简单的方法是将蒸气转化为液体并增加压力。为此,我们引入了冷凝器热交换器,位于低压汽轮机下方。在冷凝器中,冷水通过管道流动。蒸气向这个液体流中散发热量并变成了冷凝水。现在,我们可以使用泵来增加进给水的压力。通常使用多级离心泵来实现这个目的。那样,压力就会恢复到原始状态。接下来的任务是恢复温度到原始值。为此,通过锅炉向泵的出口处加热。

高装机容量的电厂通常使用一种叫作"水管锅炉"的锅炉。在锅炉内燃烧粉煤。进水最初经过一个节能站。在这里,水会从烟气中吸收能量。水通过下降管流动,然后通过水壁流动,这里它会转化为蒸气。纯净蒸气在蒸气分离器中分离。现在,工作流体恢复到了原始状态——高压高温。这个蒸气可以回馈给蒸气汽轮机,循环可以一次又一次地重复,实现持续发电。

但是,采用这种基本朗肯循环的电厂效率很低,容量也很低。我们可以通过一些简单的技术显著提高电厂的性能。在过热的情况下,即使液体已经转化为蒸气,进一步加热蒸气,使其过热。蒸气的温度越高,循环效率越高。只要记住卡诺定理最大热效率是可能的。但是,蒸气汽轮机材料不耐受超过600℃的温度,所以超热限制在该阈值内。蒸气在一排排叶片间行进的过程中温度下降。因此,提高电厂效率的有效途径是在第一级涡轮之后增加更多的热量。这就是再加热,它将再次提高蒸气的温度,从而实现更高的功率输出和更高的效率。电厂的低压侧容易吸入大气空气,即使采用了复杂的天花板设计。随着时间的推移,进给水中的溶解气体会损坏锅炉材料。为了去除这些溶解气体,引入了开式给水加热器。来自汽轮机的热蒸气被混合到进给水中。产生的蒸气泡泡会吸收溶解气体。混合还会预热进给水,这有助于更大程度地提高电厂的效率。所有这些技术使得现代电厂的效率达到40%—45%。

现在我们来看一下加热和散热是如何在实际电厂中进行的。冷却液体通过冷却塔供应到冷凝器。从冷凝器出口加热过的水在冷却塔中被喷洒出来,进而引发自然空气对流,喷洒的水失去热量,这就是为何冷凝器进口总是能提供被降温过的液体。在加热处一侧,燃烧的煤会产生许多污染物。我们不能直接将这些污染物排放到大气中。因此,在将其传输到烟囱之前,废气会在电除尘器中得到清除,电除尘器利用高电压静电板来吸收污染物颗粒。

英翻中练习

(1) Analysis of China's thermal power industry market in 2023

Thermal power generation remains the backbone of China's electricity production. Major listed companies in the industry include Huaneng International (600011), Huadian International (600027), Datang Power (601991), State Power Investment Corporation (600795), and Zhejiang Energy (600023).

—**Thermal power generation remains the backbone of China's electricity production.**

As clean energy continues to develop and penetrate the market, the proportion of thermal power generation in China's total electricity production has been declining from 2010 to 2022. In 2022, the proportion of thermal power generation dropped to 69.8%, with a year-on-year increase of 1.8 percentage points. Although the proportion of thermal power generation has been gradually decreasing, its actual power generation still surpasses half of China's total electricity production. Therefore, despite the decreasing proportion, thermal power generation remains a crucial pillar in China's power production.

—**The cumulative installed capacity of thermal power generation has been increasing year by year.**

According to the data released by the National Energy Administration, China's cumulative installed capacity of thermal power generation has been increasing year by year from 2017 to 2022. In 2022, China's cumulative installed capacity of thermal power generation reached 1.332 billion kilowatts, with a year-on-year growth of 2.70%. The growth rate has slightly decreased compared to 2021.

—**The thermal power generation capacity has been increasing year by year.**

According to the data released by the National Energy Administration, China's thermal power generation has been increasing year by year from 2017 to 2022. In 2021, China's thermal power generation reached 5.81 trillion kilowatt-hours, with a year-on-year growth of 9.01%. The growth rate significantly increased compared to 2020, mainly due to the influence of extreme weather conditions. In 2022, China's thermal power generation reached 5.85 trillion kilowatt-hours.

—**In the past two years, investment in thermal power generation projects in China has experienced rapid growth.**

From the perspective of thermal power investment in our country, the investment in thermal power generation projects in China has exhibited fluctuations from 2017 to 2022. Among them, the investment in thermal power generation projects in China experienced rapid growth from 2021 to 2022, reaching 90.9 billion yuan in 2022, a year-on-year increase of 35.27%.

—**The average utilization hours of power generation equipment in Chinese thermal power plants have shown a fluctuating downward trend.**

From 2011 to 2022, the average utilization hours of thermal power equipment in China have exhibited an overall downward trend with fluctuations. In 2022, the average utilization hours of thermal power equipment in China decreased to 4,379 hours, reflecting a year-on-year decrease of 1.55%.

(2) Generating electricity is truly a genius invention, and its principle is actually quite simple. It involves using water or steam to rotate the generator and produce electricity.

First, coal is transported by train from a coal-producing region. Through excavators and conveyor belts, the coal is transferred to the crusher, where it is crushed into fine particles and then loaded into a coal hopper. Underneath the hopper, there is a coal feeder that controls the amount of coal in the boiler. The coal particles are then sent to the grinder and continuously ground until they become fine coal powder. This is done to ensure that the coal powder can burn fully. Next, the coal powder is blown into the furnace of the boiler by a fan, where it is mixed with high-temperature air, creating intense flames. At this point, the temperature in the furnace can reach 2,200℃. The tremendous heat causes water to boil and generates a large amount of steam. This steam is supercritical, meaning it is high-temperature and high-pressure, which improves efficiency compared to ordinary steam. The steam at 1,000℃ is first sent to a high-pressure turbine, which causes the turbine shaft to rotate.

During this process, the pressure and temperature of the steam decrease significantly, so they need to be sent back to the boiler for reheating. Once the steam reaches the required temperature, it is sent to the low-pressure section of the steam turbine, where its internal energy is converted into mechanical energy.

The rotation of the turbine shaft, which can reach a speed of 3 600 revolutions per minute, drives the rotor of the generator. Through electromagnetic induction, the mechanical energy is ultimately converted into electrical energy.

After the electricity is generated, it goes through a step-up transformer. Here, the voltage is increased to several hundred thousand volts to reduce losses during transmission. Under the transmission of the power grid, the electrical energy flows into millions of households.

However, the process is not yet complete. The steam leaving the turbine condenses back into water through a cooling circulation pipeline. This water can be recycled and sent back to the boiler for further use. As for the ash produced from combustion, it either falls to the bottom of the boiler or is collected through dust collectors and can be used as raw material for cement or disposed of in landfills. The waste gases from combustion are also not released casually. They are only discharged into the atmosphere after undergoing dust removal, desulfurization, and denitrification processes.

Text 3 Mega Hydropower Stations in Southwest China

The Three Gorges Project is widely regarded as a mega project. As the world's largest hydropower station, the electric power it generates in one second is sufficient to sustain an ordinary person for more than 4 years. However, very few realize that at about 1,000 km upstream of this dam, there are two more "Three Gorges". The Jinsha River here separates the Sichuan and Yunnan Provinces. It links up 4 mega hydropower stations: Xiangjiaba Hydropower Station, the Xiluodu Hydropower Station, the Baihetan Hydropower Station, and the Wudongde Hydropower Station. Their combined installed capacity is twice that of the Three Gorges Hydropower Station.

The entire southwest China is sprinkled with mega hydropower stations like these. They stand on rivers and span between mountains. They form a series of mega hydropower bases and generate more than half of the country's hydroelectric power.

Why does the southwest region of China have so many mega hydropower stations? How did we manage to build all these colossal structures here? And how do they contribute to our daily lives?

Let us unveil the answer hidden in this extraordinary landscape.

Extraordinary Landscape

The Jinsha River Gorge is nothing short of lofty peaks and precipitous cliffs, a masterpiece created by the large river. The charging river restlessly carves the unyielding rocks into numerous narrow and treacherous gorges and reshapes the land surface. This kind of landscape can be seen everywhere in southwest China.

For billions of years plate tectonics have warped the land and formed a string of mountain ranges here. Undulating terrains act as barriers that block the water vapor coming from the oceans. Abundant precipitation and glacial meltwater merge and form large rivers, which make it the most abundant area of water resources in China. Taking the Jinsha River as an example, it has an average annual runoff of 145 billion cubic meters and a maximum flow rate that reaches 29,000 cubic meters. As this rate, it can fill up the entire West Lake in a mere 8 minutes. Apart from the abundance of water, the extreme elevation difference also gives the water a powerful push. Since the southwest region spans the first two steps of China's ladder terrain, rivers originating from snow-capped mountains in the west gush down in an unstoppable manner. For example, the Jinsha River pours down a natural elevation of more than 5,000 m over a distance of only 3,481 km.

Water abundance and huge elevation difference not only remodels the land surface, but also carries immense energy which we refer to as "hydropower resource". In southwest China, there is as much as 490,000 megawatts of theoretical hydropower reserve, which accounts for 70.6% of China's total reserve. If all this hydropower resource is converted into electric power, it will unleash extremely powerful energy. This is also why more than half of China's 13 mega hydropower bases are located in the southwest regions. These include the Jinsha River Hydropower Base.

Mega Dams

Above these extremely turbulent rivers, how do we even start to utilize this energy? First, we need to build a dam. Even though the natural elevation of Jinsha River far exceeds that of other rivers flowing on flat terrains, the power it generates over such a short distance is still way too little. Therefore, to increase the efficiency of power generation, it is necessary to erect a dam on the river to restrict water flow, which further raises the water level and heightens the elevation.

But then, building a dam on a raging river is easier said than done. Therefore, before starting the construction of the dam, we must first accomplish another huge task to temporarily divert the river and provide a dry riverbed for the construction, then restore the original river channel once the dam is completed.

In 2014, the first mega hydropower station in the lower Jinsha River was fully completed. It is the Xiangjiaba Hydropower Station. The dam of this hydropower station has a crest measuring 900 m in length, which takes 12 minutes just to walk across. The highest dam height is 162 m, which is equivalent to a 60-storey building. It has a massive body that looks similar to the famous Three Gorges Dam. Its hefty weight alone creates enough friction against the foundation to stop the coursing river. This is so-called gravity dam. It is this very dam that creates the 100 m elevation difference between the up- and downstream water levels. However, building a massive dam requires enormous amount of construction materials. For Xiangjiaba Dam, the total volume of concrete required for the dam body alone was as high as 14 million cubic meters, which can fill up more than 6,000 Olympic-size swimming pools. Simply delivering this amount of concrete already costs a fortune. So, is there any alternative?

The next dam spreads out like a white crane with its wings wide open and blocks the Jinsha River behind it. The dam's lithe figure earned itself the name "White Crane Beach" (Baihetan in Chinese). Its maximum height of 289 m makes it as tall as a 100-storey skyscraper, yet the narrowest at its crest is only 14 m wide. Despite the extremely slim body, this dam is able to withstand 16.5 million tons of thrust, which is enough to send more than 15,000 Long March-5 rockets to the sky. How is this possible? Perhaps a change in perspective is all we need to know the answer.

Looking from above, we will see that Baihetan Dam adopts an arch structure, commonly known as arch dam. This structure can transfer the thrust of the water flow to the mountains on both sides. With mountains sharing the load, arch dams do not need to be as massive as gravity dams to remain stable.

With a dam height almost twice that of Xiangjiaba Dam, the total volume of concrete needed to build Baihetan Dam was only around 8 million cubic meters, which is less than 60% of that for the former. Moreover, as a statically indeterminate structure, arch dams exhibit extremely high levels of safety that their overload capacity can be up to more than 10 times the design load. At the same time, arch dams have been growing taller and taller. Worldwide, half of the dams taller than 200 m are now arch dams. But not every location is suited to hold this type of dam. Only when all the extremely harsh geological and topographical conditions are fulfilled such as precipitous and narrow gorges and hard and intact bedrock can the arch dam be tightly clipped in the gorges.

Fortunately, the numerous gorges in southwest China provide the best sites for arch dam construction. As a result, all types of arch dams have sprung up on this land. These include, the 240 m tall Ertan Dam, the 270 m tall Wudongde Dam and the tallest dam in the world today: the Jinping Grade I Hydropower Station Dam which is 305 m tall. As these gigantic dams cluster around the southwest, even the feistiest rivers have to slow down their pace and build up energy here. And to further convert this energy into a usable form, we will need a mega hydropower station.

Mega Hydropower Stations

Dams are not simply an impermeable wall. Apart from the flood discharging outlets visible from outside, there are intricate pipelines on the inside that run towards the heart of the hydropower station: the hydroelectric turbine generator. Through diversion pipes, river water enters the generator set where it propels the hydraulic turbine and converts gravitational potential energy into mechanical energy. The turning turbine drives synchronized rotation of the generator, which in turn converts mechanical energy into electrical energy, thereby generating electric power. Throughout the entire process, we utilized nothing but the energy of flowing water. Since no fuel is consumed, there will not be any emission of greenhouse gas like carbon dioxide or other exhaust gas. Therefore, hydropower is a renewable clean energy.

On wide rivers, we can build a hydropower plant within the dam or behind it to accommodate these enormous generator sets. But inside the confined gorges, how can these facilities fit? The secret lies within this mountain, where there is the largest underground cavern cluster in the world at present. 246 crisscrossing tunnels turn this place into a maze with a total length approaching 217 km. In particular, the largest cavern is 438 m long, 34 m wide and 88.7 m tall, hence can easily accommodate 13 Boeing 737 aircrafts. In addition, it harbours the "heart" of the world's most powerful hydro-generator. The hydro-generator with a single-unit capacity of 1 million kilowatts is completely designed and built by the Chinese. Running at full capacity, the electricity it generates within 1 hour is comparable to the annual consumption of more than 1,000 persons.

In general, the larger the single-unit capacity of a hydro-generator, the higher the efficiency. The 1-million-killowatt hydro-generator of Baihetan Dam has an energy conversion efficiency of up to 96.7%. Almost every single drop of water has been put to use. It is difficult to imagine that until the mid-1990s, none of the domestically built hydro-generator in China had a single-unit capacity beyond 300,000 kilowatts. Yet we seized all opportunities emerging around the Three Gorges Project and finally managed to accomplish the 700,000-kilowatt breakthrough. Today, from the Xiluodu Hydropower Station with a single-unit capacity of 770,000 kilowatts to the Xiangjiaba of 800,000 kilowatts and from the Wudongde of 850,000 kilowatts to the Baihetan of 1 million kilowatts, the 4 mega hydropower stations on Jinsha River witnessed every step of China's hydro-generator advancement.

When the Baihetan Hydropower Station is fully completed, more than 62 billion kilowatt hours of clean electricity can be produced every year. By then, the clean electricity will cross mountains and gorges in the southwest and travel through the high-voltage transmission lines. It will arrive in East China 2,000 km away, and invigorate the magnificent West-to-East Power transmission network. As such, we have built all these mega hydropower stations in a place blessed with the richest hydropower resources in the world.

In China, tens of thousands of hydropower stations are operating relentlessly on restless rivers. Together, they generate more than a trillion kilowatt hours of clean electricity every year, which translates to saving 430 million tons of standard coal equivalent and reducing 1.12 billion tons of carbon dioxide emission.

For sure, there is still a long way to go. Ecological conservation, sediment control, resettlement of the locals and disaster monitoring all require attention and coordination. But one thing is certain. On the path towards emission reduction and carbon neutrality, hydropower will definitely take on an indispensable role. For thousands of years, rivers have nurtured both the land and our civilization. Today, these rivers are benefitting their people in a different way. From the Shilongba hydropower Station with an installed capacity of only 480 kilowatts to the Three Gorges Hydropower Station with an installed capacity of 22.5 million kilowatts, and further to the Baihetan Hydropower Station which has the world's largest single-unit capacity, countless grandiose dams and hydropower stations have been erected one at a time by multiple generations of builders. Because of their selfless effort, we are now able to utilize the generous gift from this land to deliver light and warmth to all in China, to support our progression toward modernization, and to create a better future full of hope.

Notes

(1) installed capacity 装机容量:指某种发电设备或能源的总装机能力。
(2) be sprinkled with 点缀着,散布着
(3) plate tectonics 板块构造理论:指地球上陆地和海洋表面的构造变动和演化。
(4) average annual runoff 年均径流:每年流出的水量平均值。
(5) ladder terrain 阶梯地形:逐渐升高或降低的地势。
(6) natural elevation 自然海拔:地理地势的高低。
(7) theoretical hydropower reserve 理论水电储备:指水电潜在资源的理论储备量。
(8) gravity dam 重力坝:一种利用坝体自身重力抵抗水压力的大坝结构。
(9) arch structure 拱形结构:一种将水流推力转移至两侧山脉的形状。
(10) statically indeterminate structure 超静定结构:指构造中存在超过必要数量的约束,提高了结构的稳定性。
(11) flood discharging outlets 泄洪口
(12) hydroelectric turbine generator 水轮发电机组:由水轮机和发电机组成的发电设备。
(13) diversion pipe 引水管道:将河水引入水轮发电机组的管道。

（14）gravitational potential energy 重力势能：物体由于位置高低差所具有的能量。
（15）mechanical energy 机械能：物体具有的动能和势能之和。
（16）synchronized rotation 同步旋转：指水轮机和发电机的旋转速度保持同步。
（17）single-unit capacity 单机容量：指单台水轮发电机的额定发电能力。
（18）West-to-East power transmission network 西电东送工程：指中国将西部地区的电力输送到东部地区的输电系统。
（19）standard coal equivalent 标准煤当量：用于衡量不同能源消耗的标准。
（20）carbon neutrality 碳中和：指达到净零碳排放的状态。

参考译文

第3课 中国西南地区的超级水电站

三峡工程是许多人心中的一项超级工程。作为全世界规模最大的水电站，它平均每秒的发电量就足以满足一个中国人4年多的生活用电需求。不过，很少有人知道，就在它上游一千多公里外还有两座"三峡"。金沙江在这里隔开四川和云南两省，依次串联起了四座超级水电站：向家坝水电站、溪洛渡水电站、白鹤滩水电站，以及乌东德水电站。它们的总装机容量就达到了三峡水电站的两倍之多。

放眼整个西南像这样的超级水电站更是数不胜数。它们屹立于江河之上，坐落在群山之间，形成了一系列规模庞大的超级水电基地，为全国提供了一半以上的水利发电量。

为什么在中国西南地区会有这么多超级水电站？我们是如何在这片土地上建造起这些庞然大物的呢？而它们又会给我们的生活带来什么呢？

让我们去这片超级山河中寻找答案吧。

超级山河

金沙江大峡谷悬崖耸立，气势巍峨。而它正是这条大江的杰作。滔滔江水日夜奔腾，将坚硬的岩石下切，雕刻出一条条狭长险峻的峡谷，重新塑造了地表的样貌。而这样的景观在中国西南可谓比比皆是。

亿万年来的板块运动扭曲大地，在这里形成了一众连绵的山脉。崇山峻岭宛如屏障拦截了来自海洋的水汽。丰富的降水加上冰川融水汇聚成了大江大河。这里也成为了中国水资源最丰富的地区。以金沙江为例，其平均年径流量就高达1450亿立方米，最大流量可达29000立方米/秒。照这个速度，只要8分钟就可以填满一个西湖。除了充沛的水量，极大的地势落差则为流水注入了强大的动力。西南地区横跨我国地势的第一二级阶梯，发源于西部雪山的河流顺势而下，一往无前，势不可挡。比如，在金沙江3481千米的长度内天然落差就超过了5000米。

这样丰富的水量、巨大的落差，不仅能够塑造地表，同时还蕴藏着巨大的能量。我们将其称作"水能资源"。在中国西南地区，水能资源理论蕴藏量高达4.9亿千瓦。占据了全国的70.6%。如果能将这些水能资源转化为电力，那将是何等澎湃？这也是为何中国的13大水电基地超过一半都在西南，其中就包括了金沙江水电基地。

超级大坝

在波涛汹涌的大江大河之上，要如何才能利用这些能量呢？首先，我们需要一座大坝，尽管金沙江的天然落差已经远远超过其他地势平缓处的河流。但在短距离内，仅凭这样的落差产生电力还远远不够。所以为了提高发电效率，我们通常会在河流中间修建挡水的大坝。进一步抬高水位，提升落差。

但是，要在汹涌的江水之中修建大坝，谈何容易。因此，早在大坝开工之前，我们就必须先完成一项浩大的工程。让河流改道，从而为大坝腾出干燥的施工场地。直到大坝建成之后再恢复原本的流向。

2014年，金沙江下游的第一座超级水电站建成，它就是向家坝水电站。这座水电站的大坝坝顶长度约900米，步行需要12分钟才能通过。最大坝高则达到了162米，相当于一座近60层的高楼。它体型魁梧，外貌与著名的三峡大坝相似，凭借其巨大的重量，便可以与地基产生足够强的摩擦力，以抵挡奔腾的江水。这就是所谓的重力坝。正是这座大坝让上下游的水位落差提高到了100米。但是庞大的体型也意味需要耗费大量的建筑材料。在向家坝，仅大坝的混凝土浇筑总量就高达1 400万立方米，足以填满6 000多个国际标准游泳池，如此庞大的用量仅材料运输就是一笔不小的开支。那么是否还有别的形式呢？

下面这座大坝宛如一只白鹤，张开双翅，将金沙江拦在身后。它体态轻盈，恰如其名。人称白鹤滩。它的最大坝高可达289米，相当于一座100层的高楼。坝顶最窄处宽度却仅有14米。如此轻薄的坝体，却可以承受1 650万吨的推力。比15 000多个长征五号火箭的起飞推力不相上下。这是如何做到的呢？也许换个视角我们就能找到答案。

从空中俯瞰，可以发现，白鹤滩大坝采用的是一种拱形结构，我们将其称为拱坝。这种结构可以将水流的推理顺势传递给两岸山体。由于山体的分担恐怕并不需要重力坝那样庞大的体量便足以维持自身的稳定。

尽管坝高接近向家坝大坝的两倍，但白鹤滩大坝所需的混凝土却只有约八百万立方米，不到前者的60%。此外，作为一种超静定结构，拱坝具备极高的安全性，超载能力甚至可以达到设计指标的十倍以上。与此同时，拱坝也越修越高，在全球200米以上的高坝中占据了半壁江山。但并不是哪里都可以建造这样的拱坝。只有满足极其苛刻的地质地形条件，比如高陡狭窄的河谷，坚硬完整的基岩等等，才可以将拱坝稳稳地卡在河谷之中。

幸运的是，西南地区的众多峡谷恰好提供了修建拱坝的绝佳地址。于是各式各样的拱坝都在这片土地上应运而生。比如240米高的二滩大坝，270米高的乌东德大坝，以及目前世界最高坝，305米高的锦屏一级水电站大坝。当这些高坝大库在西南大地集结，原本桀骜不驯的江河，也不得不暂时收敛它们的脚步，在这里积攒能量。然而，要将这些能量进一步转化，我们还需要一座超级电厂。

超级水电站

大坝并不是一堵密不透风的挡水墙。除了可以看见的泄洪孔，其内部还拥有复杂的管道，而它们将通向水电站的心脏——水轮发电机组。江水顺着饮水管到达机组，冲击水轮机叶片旋转，从而将水流的重力势能转化为机械能。水轮机同步带动发电机旋转，再将机械能转化为电能，最终完成电力的生产。在这个过程中，我们只是借助了流水的力量，并不消耗任何燃料，自然也不会排放二氧化碳等温室气体以及其他废气，所以水电还是一种可再生的清洁能源。

对于河道宽阔的地方,我们可以直接在大坝中间或者后方修建厂房,以安置这些庞大的机组。但是如果在狭窄的山谷里,这些设备又该安放于何处?秘密就在这座大山里。这座大山里有现在世界上规模最大的地下洞室群,246条隧洞纵横交错,宛若迷宫。总长可达217公里。其中最大洞室长438米,宽34米,高88.7米,甚至可以轻松停下13架波音737客机。不仅如此,这里还安放着世界上最强的水电"心脏",单机容量百万千瓦的水轮发电机组。它完全由中国人自行设计建造,一台这样的机组只要全力运转1小时,就足以满足千余人全年的用电。

一般来讲,水轮发电机组的单机容量越大,效率也就越高。白鹤滩的百万千瓦水轮发电机组能量转换效率最高可达96.7%,几乎可以说把每一滴水都用到了极致。很难想象,就在上世纪90年代中期,我国自主建造的水轮发电机组最大单机容量却仅有30万千瓦。借着三峡工程的契机,我们才终于实践了70万千瓦的大跨越。今天从西洛渡水电站的77万千瓦,到向家坝水电站的80万千瓦,从乌东德水电站的85万千瓦,再到白鹤滩水电站的100万千瓦,金沙江上的这四座超级水电站也见证了中国水电机组的升级之路。

当白鹤滩水电站全面建成之后,每年可以产生超过620亿千瓦时的清洁电力。届时,它们将跨越西南的高山峡谷,通过各级高压输电线路,送往近2 000公里外的华东地区,为西电东送这张壮丽的网络注入新的能量。就这样,我们在全世界水能资源最富足的地方,修建起了如今这些超级水电站。

放眼全国,伫立江河之上的数万座水电站与奔流不息的江水一样,昼夜不停的运转,每年可提供超过1万亿千瓦时的清洁电力,相当于节约标准煤4.3亿吨,减排二氧化碳11.2亿吨。

当然,未来的路还很长。水电建设的生态保护,泥沙治理。移民安置,灾害监测等也需要统筹兼顾。但可以肯定的是,在减少碳排放,实现碳中和的道路上,水电必将承担起至关重要的使命。滔滔江水奔腾了千万年,滋润大地,哺育文明。而今天这些大江大河正在用另一种方式造福它的子民。从装机容量仅有480千瓦的石龙坝水电,到2 250万千瓦的三峡水电站,再到单机容量世界第一的白鹤滩水电站,一座又一座大坝、一个又一个电厂,在一代又一代建设者的脚下拔地而起。也正因如此,今天的我们才能够利用这片山河的馈赠,去点亮中国大地上的万家灯火,去支撑现代化的滚滚浪潮,去创造一个充满希望的美好明天。

英翻中练习

(1) Hydropower works by utilizing the energy of water to rotate a turbine, which in turn drives a generator to produce electricity. The principle of hydropower generation is as follows: dams are built on rivers with a significant water level difference. A large amount of water is stored in reservoirs behind the dams, and the intake is located at the bottom of the dam wall. Gravity forces the water to flow through the penstock inside the dam, pushing the spiral fan blades of the turbine to rotate. The rotation of the turbine shaft drives the generator to produce electricity. The generated electricity is then transmitted to homes and factories through power transmission lines connected to the generator. The water continues to flow past the turbine blades and is eventually released

into the river through the tailrace channel.

Engineers explain the working principle of generators as follows: the turbine converts the energy of flowing water into mechanical energy, and the hydroelectric generator converts this mechanical energy into electrical energy. The operation of the generator is based on the principle discovered by Faraday, which is electromagnetic induction. He found that a magnetic field passing through a conductor can induce an electric current. In large generators, a set of electromagnetic coils made up of layers of magnetic iron pieces are wound in a circular path. This device is called the magnetic pole, which is installed on the outer periphery of the rotor. The rotor is connected to the turbine shaft and rotates at a constant speed. The rotation of the rotor causes the magnetic poles (electromagnets) to induce movement in the conductors installed on the stator, resulting in the generation of electric current and voltage at the output terminals of the generator.

(2) A water turbine is a power machine that converts the energy of flowing water into rotational mechanical energy. It belongs to the turbomachinery category in fluid mechanics. As early as around 100 BC, China had the prototype of a water turbine called a water wheel, which was used for irrigation and driving grain processing equipment. Modern water turbines are mostly installed in hydropower stations to drive generators for electricity generation. Water turbines and auxiliary equipment are important components of hydropower facilities and are essential equipment to make full use of clean renewable energy to achieve energy-saving, emission reduction, and environmental pollution reduction. The development of water turbine technology is in line with the development scale of China's hydropower industry.

Water turbines can be divided into two major types based on their working principles: impulse turbines and reaction turbines. Impulse turbines rotate mainly due to the impact of water flow, with the pressure of water flow remaining constant during operation. It mainly converts kinetic energy. Reaction turbines rotate due to the reaction force of water flow, and both the pressure and kinetic energy of water flow change during operation, with the main conversion being of pressure energy.

Driven by the strong demand for electricity in China, the manufacturing industry of water turbines and auxiliary equipment has entered a period of rapid development, with significant improvement in economic scale and technological level. China's water turbine manufacturing technology has reached an advanced level in the world.

Text 4 What is Nuclear Energy?

Nuclear Explained

Nuclear energy is a form of energy released from the nucleus, the core of atoms, made up of protons and neutrons. This source of energy can be produced in two ways: fission—when nuclei of atoms split into several parts—or fusion—when nuclei fuse together.

The nuclear energy harnessed around the world today to produce electricity is through nuclear fission, while technology to generate electricity from fusion is at the R&D phase.

What is nuclear fission?

Nuclear fission is a reaction where the nucleus of an atom splits into two or more smaller nuclei, while releasing energy. For instance, when hit by a neutron, the nucleus of an atom of uranium-235 splits into two smaller nuclei, for example, a barium nucleus and a krypton nucleus and two or three neutrons. These extra neutrons will hit other surrounding uranium-235 atoms, which will also split and generate additional neutrons in a multiplying effect, thus generating a chain reaction in a fraction of a second.

Each time the reaction occurs, there is a release of energy in the form of heat and radiation. The heat can be converted into electricity in a nuclear power plant, similarly to how heat from fossil fuels such as coal, gas and oil is used to generate electricity.

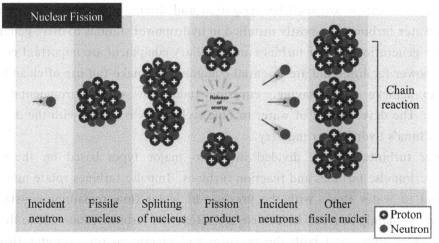

Nuclear fission (Graphic: A. Vargas/IAEA)

How does a nuclear power plant work?

Inside nuclear power plants, nuclear reactors and their equipment contain and control the chain reactions, most commonly fuelled by uranium-235, to produce heat through fission. The heat warms the reactor's cooling agent, typically water, to produce steam. The steam is then channelled to spin turbines, activating an electric generator to create low-carbon electricity.

Mining, enrichment and disposal of uranium

Uranium is a metal that can be found in rocks all over the world. Uranium has

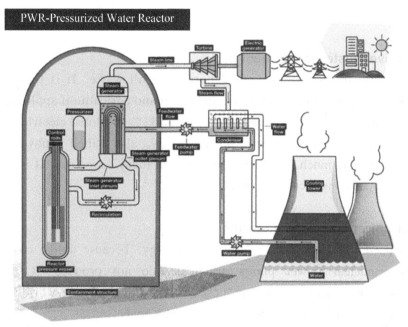

Pressurized water reactors are the most used in the world. (Graphic: A. Vargas/IAEA)

several naturally occurring isotopes, which are forms of an element differing in mass and physical properties but with the same chemical properties. Uranium has two primordial isotopes: uranium-238 and uranium-235. Uranium-238 makes up the majority of the uranium in the world but cannot produce a fission chain reaction, while uranium-235 can be used to produce energy by fission but constitutes less than 1 per cent of the world's uranium.

To make natural uranium more likely to undergo fission, it is necessary to increase the amount of uranium-235 in a given sample through a process called uranium enrichment. Once the uranium is enriched, it can be used effectively as nuclear fuel in power plants for three to five years, after which it is still radioactive and has to be disposed of following stringent guidelines to protect people and the environment. Used fuel, also referred to as spent fuel, can also be recycled into other types of fuel for use as new fuel in special nuclear power plants.

What is the Nuclear Fuel Cycle?

The nuclear fuel cycle is an industrial process involving various steps to produce electricity from uranium in nuclear power reactors. The cycle starts with the mining of uranium and ends with the disposal of nuclear waste.

Nuclear waste

The operation of nuclear power plants produces waste with varying levels of radioactivity. These are managed differently depending on their level of radioactivity and purpose.

Radioactive Waste Management

Radioactive waste makes up a small portion of all waste. It is the by-product of millions of medical procedures each year, industrial and agricultural applications that use radiation and nuclear reactors that generate around 11% of global electricity.

The next generation of nuclear power plants, also called innovative advanced reactors, will generate much less nuclear waste than today's reactors. It is expected that they could be under construction by 2030.

Nuclear power and climate change

Nuclear power is a low-carbon source of energy, because unlike coal, oil or gas power plants, nuclear power plants practically do not produce CO_2 during their operation. Nuclear reactors generate close to one-third of the world's carbon-free electricity and are crucial in meeting climate change goals.

Notes

（1）fission 裂变：原子核分裂成两个或更多较小的核，释放能量的反应。

（2）fusion 聚变：原子核融合在一起的过程，释放能量的反应。

（3）nucleus 原子核：原子的中心部分，由质子和中子组成。

（4）neutron 中子：原子的一种粒子，没有电荷。

（5）multiplying effect 倍增效应：当额外的中子进入裂变反应中，会导致更多的原子核继续裂变，反应逐渐放大。

（6）nuclear reactor 核反应堆：用于控制核裂变链式反应的设备。

（7）cooling agent 冷却介质：用于从反应堆中带走热量的物质，通常是水。

（8）pressurized water reactor 压水堆反应堆：当前世界上最常用的核反应堆类型之一。

（9）isotope 同位素：具有相同化学特性但质量和物理性质不同的元素形式。

（10）uranium enrichment 铀浓缩：通过提高样品中铀-235 的含量，使天然铀更容易发生裂变的过程。

（11）spent fuel 泛燃料：在核反应中燃烧过的核燃料，不能再继续使用的燃料。

（12）innovative advanced reactors 革新型先进堆、创新先进反应堆：下一代核电厂，相比今天的反应堆将产生更少的核废料。

参考译文

第 4 课　什么是核能？

解读"核"

核能是从质子和中子组成的原子核（原子的核心）释放的一种形式的能量。这种能源可以通过两种方式产生：裂变——原子核分裂成若干部分；或聚变——原子核融合在一起。

当前，世界各地用于生产电力的核能是通过核裂变产生的，而利用核聚变生产电力的技术正处于研发阶段。

什么是核裂变？

核裂变是一个原子核分裂成两个或多个更小的原子核并释放能量的一种核反应。例如，当一个铀-235原子核被一个中子撞击时，分裂成两个更小的原子核，如一个钡原子核、一个氪原子核和两个或三个中子。这些额外的中子将撞击周围其他铀-235原子，这些铀-235原子也将以倍增效应分裂并产生额外的中子，从而在瞬间产生链式反应。

每次反应发生时，都有热和辐射形式的能量释放。释放的热可以在核电厂被转化为电，如同煤、天然气和石油等化石燃料产生的热被用于发电那样。

核电厂是如何工作的？

在核电厂内部，核反应堆及其设备包容并控制以最常用的铀-235作为燃料的链式反应，通过裂变产生热。所产生的热使反应堆冷却剂（通常是水）升温，产生蒸气。蒸气随即被导入旋转涡轮机，驱动发电机生产低碳电力。

铀的开采、浓缩和处置

铀是存在于世界各地岩石中的一种金属。铀有几种天然存在的同位素，同位素是一种元素的不同形式，它们的质量和物理性质不同，但化学性质相同。铀有两种原生同位素：铀-238和铀-235。铀-238在全世界铀中占大多数，但它不能产生裂变链式反应，而铀-235可通过裂变产生能量，但它在全世界铀中占比不到1%。

为了使天然铀更容易发生裂变，有必要通过一个称为铀浓缩的工艺来增加给定样品中铀-235的含量。铀经过浓缩，可以在核电厂作为核燃料有效地使用三至五年，此后它仍然具有放射性，因此必须按照严格准则进行处置以保护人类和环境。使用过的燃料也被称为乏燃料，还可以被回收制成其他类型的燃料，在特殊核电厂中作为新燃料使用。

什么是核燃料循环？

核燃料循环是通过多个步骤在核动力堆中利用铀生产电力的工业过程。该循环始于铀矿开采而终于核废物处置。

核废物

核电厂的运行产生放射性水平不同的废物。根据它们的放射性水平和用途，以不同的方式对其进行管理。

放射性废物管理

放射性废物占所有废物的一小部分。它是每年数以百万计医疗程序、利用辐射的工业和农业应用以及发电量占全球电力约11%的核反应堆的副产品。

下一代核电厂也被称为革新型先进堆，将产生比当前的反应堆少得多的核废物。预计到2030年,这种先进堆可能将进行建设。

核电与气候变化

核电是一种低碳能源,因为与煤、石油或天然气发电厂不同,核电厂实际上在运行中不产生二氧化碳。核反应堆生产了全球近三分之一的无碳电力,对实现气候变化目标至关重要。

> 英翻中练习

(1) On Thursday, Rafael Mariano Grossi, IAEA Director General, visited Shidaowan Nuclear Power Plant, which is home to a 200 MWe high-temperature gas-cooled reactor.

Around the world, about 60 nuclear power reactors are under construction, and more than 400 are in operation. While China comes in third for nuclear power generation today—after France and the United States of America—by 2030, the country is expected to lead the world in installed capacity of nuclear power. "The speed of China's growth of nuclear power is remarkable, from the first nuclear reactor that connected to the grid in 1991 to the 55 nuclear reactors in operation today," said Rafael Mariano Grossi, IAEA Director General. MrGrossi was in China this week and met with several high-level officials and visited nuclear facilities and institutions in Beijing, Shandong and Shanghai.

China is ranked first for the number of nuclear power reactors under construction; it has 22 reactor units under construction, concentrated along the east and south coasts. By 2035, China's nuclear power generation will account for 10 per cent of the country's electricity generation, according to the latest Blue Book of China Nuclear Energy Development Report.

"China is a leader in the promotion of the peaceful uses of nuclear energy in terms of installed capacity and technology. This impressive development is showcased by Shidaowan Nuclear Power Plant," MrGrossi said, during his visit to Shidaowan, also known as Shidao Bay, in Shandong Province in eastern China on Thursday.

"Shidaowan's 200 MWe high-temperature gas-cooled reactor (HTGR) has unique design features that make it more efficient and inherently safe, and it has great potential to help meet net zero goals," MrGrossi added. The HTGR first connected to the grid in December 2021. Construction of two CAP1400 pressurised water reactor units is also under way at Shidaowan.

(2) China's nuclear industry dates back to 1955. After decades of development, China has established a complete system of nuclear science, technology and industry, evolving from having no nuclear power to owning homegrown Gen III technology, achieving technology upgrading and expanded production of nuclear fuel, and realizing rapid development and industrial application of nuclear technology. On the occasion of the 60th anniversary of China's nuclear industry in January 2015, President Xi Jinping

commented, "Over the past 60 years, several generations of scientists and others in the nuclear industry have made outstanding contributions to our national security and economic development. Their hard work, innovation, and trailblazing efforts have enabled our nuclear industry to develop from scratch and thrive with remarkable achievements. As a high-tech strategic industry, the nuclear industry is an important cornerstone of our national security." He went on to emphasize "the need for a continued focus on safety and innovation in pursuit of peaceful uses of nuclear energy, and the importance of further improving the competitiveness of our nuclear industry in an effort to add a new great chapter to the annals of the development of the industry".

Text 5　Wind Power Generation

Wind, the wind from the ocean.

If all the available offshore wind in the world were to be converted into electricity, the electricity generated during the same period could be more than 20 times of all current man-made power plants. However, compared to the vast sea, humans are far too small. To collect such a huge volume of energy, we have only one approach: to build our own giants.

The largest wind turbines on the Bohai Sea stands at 233 meters high, which is equal to a building of nearly 80 stories. However, this does not include its part under the sea. The impeller has a diameter of 220 meters, which can sweep through an area in rotation that is equal to the area of 5.3 standard football fields. The power generated by only one rotation is sufficient for a family of three to use for two days. And its weight is more than 900 tons, which is heavier than China's most potent carrier rocket, the Long March 5.

So, how does such a giant manage to keep itself steady in the wind and waves on the sea? And in the future, where will it go?

The Birth of a Giant

Let's go to the birthplace of the giants to find it out.

(Wind power equipment manufacturing base, Rudong of Jiangsu Province)

You will see one of the most important components of this "giant". They are produced in this factory, and then one by one, delivered out of a 6-meter-high, 6-meter-wide gate. Yes, one piece at a time, because it's too big.

A blade of an offshore wind turbine has a length of 110 meters that is nearly equal to 200 adults standing in a row. But you can even hardly imagine that, today the longest turbine blade on the Chinese sea is 123 meters, which is 20 meters longer than the largest turbine blade on land. Such a huge structure is not made of metal, but a kind of synthetic material composed of resin and glass fiber. It is this material that makes the blades light enough, while still strong enough. This is also the only way to facilitate offshore construction while allowing the blades to withstand the unruly offshore wind for at least 20 years.

The shell of a blade is 110 meters long. The production of such a large blade is not a one-off process, but requires the use of special moulds to produce the upper shell, lower shell and core beam separately, and then put them together as a whole. What was also born in these factories is the body of the giant. The wind turbine tower whose diameter can be more than 8 meters and height more than 100 meters. Its entire body is so tall that even in a factory that covers an area of 5 football fields, it can only be manufactured in parts.

After completing a few months of testing, these huge components will be transported by special trucks to a special dock, where they shall wait. The transport of these components is like a silent marching ceremony. Later, they will be transported on special ships or wind power installation platforms to their "battlegrounds" above the sea.

Today, the installation of these wind turbines can be completed at one go, including transportation, navigation, lifting, hoisting, positioning, etc., almost all of the construction process. The most common one is this strange look of the wind power installation platform. It has a straight pile at each corner, which will only descend and insert into the sea floor when it reaches its planned installation location. They keep lifting until the whole platform is above the sea, so that the effects of wind and waves can be avoided. After this, the huge main crane is able to precisely align the impeller and nacelle with a mere millimetre error, despite the disruption of strong winds at high altitude. This operation can even complete the lifting of equipment as heavy as 3,000 tons at one go, that is equal to nearly 1,800 cars. Finally, the foundation, tower, blades, hubs, etc., are precisely assembled.

As the impeller starts to turn, the giant wakes. Though these huge impellers can only rotate 17 times per minute, with the gearing device in the nacelle, the speed of the generator can be up to 1 800 rpm. By now, the giant standing in the wind will take the roaring energy into its embrace. The wind that humans once could not grasp has thus become electricity, flowing to cities across the country.

However, the power of only one turbine is rather limited. It is far from enough to capture the wind of the entire ocean. However, this giant is not alone. What's next to it is a powerful legion of wind turbines.

A Legion of Giants

(Laizhou, Shandong Province, Bohai Sea)

There are 38 "sea giants" stationed here, and together they formed one of the many "giants legions" on the Bohai Sea. If we bring up the camera, we will find the mystery of their formation. These turbines are arranged in regular rows on the sea, and each row is parallel to the other, so that all turbines are oriented perpendicular to the direction of the wind in the area, thus maximizing the efficiency of power generation. The distance between the turbines is exactly 7 - 8 times the diameter of the impeller, so that the weakened and disturbed wind can be re-enhanced and do not affect other turbines. Moreover, these huge impellers are not stationary. Instead they can rotate 360° according to the season and wind direction to always receive the roaring wind head on. In the event of typhoons, they will turn to minimize the surface against the wind and stop rotating, thereby protecting themselves.

(Weifang, Shandong Province, Bohai Sea)

The power is first converged via submarine cables to the offshore booster station, and then sent to the onshore centralized control center via a high voltage electricity with much less loss, and finally into the transmission network. If we look into the whole China, by the end of 2021, there have been 5,237 offshore wind turbines installed guarding like giants on the coastlines of China. Together they made five major offshore wind power in total: Shangdong Peninsula, Yangtze River Delta, southern Fujian, eastern Guangdong and Beibu Gulf. At the same time, with 26 million kilowatts of installed capacity, China has surpassed the United Kingdom to become the country with the largest installed capacity of offshore wind power. Compared to the inland wind energy legions, the offshore wind is much less affected by the landscape and is available for more than 1.5 times longer than that of land per year.

(Jiazi Offshore Wind Farm, Shangwei, Guangdong Province)

(Nanridao Offshore Wind Farm, Putian, Fujian Province)

(Laizhou Offshore Wind Farm, Yantai, Shandong Province)

The construction of wind energy plants does not take up the scarce land resources, nor does it affect people's daily lives with noise. Plus, they are generally close to coastal areas, so there is no need to build ultra-long-distance transmission lines, which can quickly meet the demand of electricity load centers. It is predictable that as the technology of offshore wind power gradually matures, the cost will gradually decrease. In the future, we will build more and larger offshore giants to form a more powerful offshore legion, and to discover new areas where wind energy resources are more abundant.

The Expedition of Giants

Deep sea has more powerful and sustainable wind. The exploitable wind energy reserves of the sea are 1.7 times more than the offshore and 5 times more than land.

(Weifang, Shandong Province, Bohai Sea)

Our giants on the sea are bound to start an expedition to the deep sea. However, the road to the deep sea is also bound to be difficult. One of the biggest challenges is to make these huge giants stand securely in more than 50 meters underwater.

If you are in the neritic zone, when you dive underwater, you will see that what helps these huge things stand up in the wind and waves are steel pipes and conduit frames, which we call foundations. Compared to the part above the sea, these foundations below the sea does not seem to be very large, but when we penetrate through the mud, sand and rocks, we find that some of the foundation structure has long been deep underground. For this reason, they can support the body of these giants through the wind and waves and remain unshaken.

However, for the deep sea, this kind of structure is still too large, and it is much less cost-effective. People must bring novel ideas to this legion of giants. For example, giving up on the giant foundation, and replace it with a floating base. This is the floating wind turbine, the base of which is no longer plunged straight into the sea floor, but only half submerged and connected to the seabed using a few anchor cables. The base itself is in a Y-shaped structure, and the wind turbine stands on one corner of it. Through precise calculation, the gravity, buoyancy, tension, wind thrust, etc. of this structure can all balance each other. It can control the swaying angle and maintain stability in the waves, as well as resist typhoons of up to 17 levels.

Of course, in the expedition to the deep sea, we have just begun. In the future, larger blades, more powerful engines, more resistant foundations, and more robust cables are all waiting for people to break through. In a certain sense, this is an expedition of mankind itself.

Not far away in Zhanjiang, China's first deep-sea floating wind turbine is silently standing in the water at a depth of 65 meters. As the vanguard of the first expedition, people have given it a special name, Fuyao. "One day the roc will fly with the wind straight up the nine clouds." Just like one day in the future, when the wind sweeps across the deep sea, it comes the time of raging energy among thousands of impellers.

Yes, human beings are small, but these giants of the sea built by our own hands will go farther to the sea for us, so that the wind roaring for billions of years will no longer be blocked by mountains and sea. And instead, it will go over the mountains in another form to illuminate the unique light on this planet.

Notes

(1) Long March 5 长征五号：中国最强大的运载火箭型号之一。

(2) offshore wind turbine 海上风力发电机：用于在海上获得风能并转换为电能的设备。

(3) upper shell 上壳：风机叶片的上半部分外壳。

(4) core beam 芯梁、核心梁：风机叶片的中间支撑梁。

(5) wind turbine tower 风机塔、风力发电塔：用于支撑风机的塔型结构，通常高度超过100米，直径超过8米。

(6) wind power installation platform 风电安装平台：用于安装风力发电设备的平台，

通常具有直立桩用于插入海底以避免风浪的影响。

(7) hoisting 吊装:使用起重设备将重物提升至空中的过程。

(8) positioning 定位:将物体放置到预定位置的过程。

(9) nacelle 转向箱、机舱:风力发电机顶部的外壳,内部包含发电机和其他关键组件。

(10) millimetre error 毫米误差:在无风浪干扰的条件下,风力发电机精确安装时所允许的最大误差。

(11) hub 轮毂:连接风机叶片和叶轮的部件。

(12) station 驻扎:指风力发电机在海上的安置和停留。

(13) offshore booster station 近海升压站:将海上风电场的电力转换成高电压电力并输送到岸上的设施。

(14) centralized control center 集中控制中心:用于集中控制和监控风电场运行的中心设施。

(15) neritic zone 近浅海、海岸浅海带:位于海岸线附近,水深较浅的海洋区域。

(16) steel pipes and conduit frames 钢管和导管框架:用于构建风力发电塔基的钢制管道和构架。

(17) cost-effective 成本效益、性价比

(18) floating base 漂浮式基础结构、浮式基础:指安装在水上悬浮的基础结构。

(19) anchor cable 锚索、锚链:用于连接浮动风力涡轮机基础和海床,提供稳定性和支撑力的钢索。

(20) Y-shaped structure Y字型结构:具有Y形状的结构,用于支撑和平衡浮动风力涡轮机的动力和重力。

(21) swaying angle 摇晃角度、摇摆角度:指浮动风力涡轮机在波浪中摇摆的角度。

(22) roc 鹏:传说中巨大的鸟类,象征着巨大的力量和能力。

参考译文

第5课 风力发电

风,来自海上的风。

将全球海上可利用的风能全部转化为电能,同一时间内的发电量将达到当前人类所有发电厂的20多倍。可是相比浩瀚的海。人类太过于渺小。想要收集如此巨大的能量。我们只有一个方法:造出自己的巨人。

渤海上最大的风力发电机高度达到233米,相当于近80层的高楼,这还不算它的海下部分。叶轮的直径为220米,旋转时可以扫过的面积相当于5.3个标准足球场的面积。它转动一圈的发电量,就可以让三口之家用上两天,它的重量更是高达900多吨,甚至超过中国最强大的运载火箭长征五号。

身躯如此庞大的海上巨人是如何让自己稳稳站在风浪之中的?未来它又将走向何方?

巨人的诞生

让我们前往巨人诞生的地方一探究竟。

（江苏如东风电设备制造基地）

你将会看到最重要的组成部分之一。它就在这家工厂中完成生产，一次一片的从宽六米，高六米的大门运出。你没有听错，一次只会运出一片，因为它实在太大了。

这就是海上风机的叶片，110米的长度，相当于近200个成年人站成一排。但你更难以想象，今天中国海面上的风机叶片最长已达到123米，比陆地上最大的风机叶片还要长20多米。如此庞大的结构却并非用金属制成，而是由一种树脂和玻璃纤维合成的材料制造。正是这样的材料让叶片的重量够轻，但强度依然够高。也只有这样，才能在便于海上施工的同时，能让叶片和桀骜不驯的海风抗衡至少20年之久。

叶片的外壳长110米。生产如此庞大叶片并非一次成型，而是需要使用专用模具分别生产上壳、下壳和芯梁，然后将它们组合成一个整体。同样在这些厂房里诞生的还有海上巨人的身躯。风机塔直径达到八米多，高100多米。由于它实在太过于高大，因此，即便在占地面积有五个足球场大的厂房里，也只能分段制造。

经过最多长达几个月的测试后，这些巨大的组件将通过专门的运输车运送至专门的码头，并在那里静静等待。这些组件的运输像一场无声的出征仪式。而接下来它们还会换乘专门的运输船或者风电安装平台，奔赴它们在大海之上的战场。

如今，这些风电安装设备已经能一口气完成运输、航行、升降、吊装、定位等几乎所有的建造流程。其中最常见的便是外形奇特的风电安装平台。它的每个角都有一个直桩。只有抵达预定的安装位置时，这些直桩才会降下插入海底，然后不停往上顶，直到将整个平台从海面上顶起，从而避免风浪等造成的影响。而在这之后，巨大的主吊机即便在高空遭遇大风干扰，却依然能够以仅仅毫米级的误差将叶轮和机舱精准对位。甚至可以一次完成最重达3000吨的装备吊装。相当于能一次吊起1800辆轿车。最后，基础、叶片、轮毂等精密地完成组合。

叶轮开始转动，巨人随之苏醒。尽管这些庞大的叶轮一分钟才能转到17圈。但经过机舱中的变速装置发电机转速仍可达到1800转每分钟。至此，迎风而立的巨人将呼啸的能量揽入自己的怀抱。人类曾经抓不住的风也就此化作电力向全国各地的城市流淌。

不过仅仅一座风机的力量相当有限，还远不足以抓住这整片海域的海风。但是这位巨人并不孤独，因为在它的身旁是一支强大的军团。

巨人军团

（渤海山东莱州）

从面前的这片海域上，就有38位海上巨人镇守于此。它们共同组成了渤海上众多的巨人军团之一。而当我们将视角拉高，就会发现它们排兵布阵的奥秘。这些风机在海面上十分规则地排成数行，且每一行都彼此平行，这是为了让所有风机的朝向基本垂直于当地的主风向，从而让发电效率最大化。而风机之间又恰巧留出了相当于七到八倍叶轮直径的距离，这样一来，能让受到风机削弱和扰动的海风重新得到增强，避免影响其他风机的正常运转。这些庞大的叶轮也并非固定不动，而是随着季节和风向360度旋转，时刻保持能正面迎接呼啸的海风。而一旦遭遇台风则转而将迎风面降到最小，并停止转动，以此来保护自己。

（渤海山东潍坊）

电力先通过海底电缆汇聚到海上升压站，再通过损耗小得多的高压电送到陆上集中控制中心,最后进入输电网。放眼全中国,截至2021年底,已有5 237座海上风机,如同巨人一般镇守在我国的海岸线上,共同组成了山东半岛、长三角、闽南、粤东和北部湾共计五大海上风电基地。同时也凭借2 600万1 000瓦的装机容量让中国一举超越英国,成为全球海上风电装机规模最大的国家。而相比于内陆的风电军团,海风受地形影响更小,每年的可利用时长是陆地的1.5倍以上。

(广东汕尾甲子海上风电场)
(福建莆田南日岛海上风电场)
(山东烟台莱州海上风电场)

在这里建设风电厂既不占有稀缺的土地资源,也不会因噪声影响人们的日常生活。再加上海上风电厂的位置,一般紧邻沿海地区,因此,不需要再修建超远距离的输电线路,就能迅速满足用电负荷中心的需求。可以预测的是,随着海上风电的技术逐渐成熟,成本逐渐降低,未来我们还会造出更多更大的海上巨无霸,组成更强大的海上军团,前往风能资源更加丰沛的地方开辟新的战场!

巨人的远征

深海具有更持久的海风!储量是近海的1.7倍,陆地的五倍。
(渤海山东潍坊)

我们的海上巨人势必要开启一场前往深海的远征。而走向深海的道路也势必困难重重。最大的挑战之一莫过于如何让这些身躯庞大的巨人在超过50米的海水中,依然能够站稳脚跟。

如果在近浅海,当你潜入水下就会发现,帮助这些庞然大物停立在风浪之中的是钢铁打造的钢管或者导管架,我们称之为基础,相比于水面之上的部分,这些水面之下的基础看上去似乎也并没有多么庞大,但当我们穿透泥沙和岩石就会发现,其中一些基础的结构早已深深插入地下。也正因如此,才能支撑这些巨人的身躯,历经大风大浪依然毫不动摇。

然而在深海海域,这种结构的基础将变得过于庞大,性价比也将大打折扣。人们必须为这支巨人军团带来新的思路,如干脆放弃固定式的巨大基座,改用水上漂的基座。这就是漂浮式风机,它的基座不再直插海底,而是仅有一半潜入水下,并通过几根锚索与海底相连。基座则是Y字形结构,风机立于其中一角之上。通过精妙的计算,这种结构的重力、浮力、张力和风推力等均可以相互平衡,不仅能控制摇晃角度,在海浪中保持稳定,甚至还能抵抗最高17级的台风。

在进军深海的道路上,我们才刚刚开始,未来更大尺寸的叶片、更大功率的发电机、更抗风浪的基础、更长更牢固的电缆,都在等待着人们去突破。从某种意义上讲,这是一场人类自己的远征。

不远处的湛江,中国首台深海漂浮式风力涡轮机静静地矗立在65米深的水中。作为率先出征的前锋,人们给了它一个特别的名字——扶摇号。"大鹏一日同风起,扶摇直上九万里。"恰如有朝一日,当遥远的海面上风起云涌之时,也正是成千上万个叶轮间能量汹涌之时。

人类纵然渺小,但是我们打造的海上巨人将会代替我们去到更远的大海,让咆哮了亿万年的海风不再因山海而阻隔,而是以另一种形式翻山越岭点亮这颗星球上独一无二的光芒。

英翻中练习

(1) Wind. It's the largest renewable energy source in the US. In one month, the average wind turbine can generate enough power for more than 940 American homes.

Here's how wind energy works, plus its upsides and downsides. Wind energy is one of the few renewable, clean energy sources. We say "renewable" because the supply of wind is basically limitless and naturally replenishable, and it's "clean" because no pollutants are released in the process. While fans use electricity to make wind, wind turbines use wind to make electricity. That wind can be used directly like for grinding grain or pumping water, or indirectly through a generator, which can convert the wind's power into electricity. Wind turns blades around a rotor or hub, which spins a generator, which creates electricity. The electricity then travels through underground cables to a substation, which connects to an electric grid, which then connects to, say, your home.

The taller the turbine, and the wider the blades, the more wind gets captured and the more energy gets created. There are two main types of commercial wind turbines: those on land and those at sea. Offshore winds tend to be stronger during the day, providing energy when consumer demand is highest. Most land-based wind turbines are stronger at night when electricity demands are lower.

(2) The best places to build wind farms are on the coast, the Midwest and Great Lakes area. That's where the highest speed winds are. As of right now, almost all US wind farms are on shore. In 94 minutes, an average wind turbine generates enough energy to power an average US home for a month.

One downside of the wind industry: its potential to harm wild animals, mainly birds and bats. The impact can be direct through collisions and indirect through noise pollution and possible habitat loss.

In addition to sustainability, another upside: cost. In 2021, the cost of producing a megawatt hour of electricity from a new wind turbine was $26 to $50, while the same amount of electricity from the cheapest type of natural gas plant ranged from $45 to $74. Wind turbines, which accounted for more than 9% of electricity in the US in 2021, last roughly 20 to 25 years before they must be replaced.

Text 6　How do solar panels work?

The Earth intercepts a lot of solar power: 173 thousand terawatts. That's ten thousand times more power than the planet's population uses. So is it possible that one day the world could be completely reliant on solar energy?

To answer that question, we first need to examine how solar panels convert solar energy to electrical energy. Solar panels are made up of smaller units called solar cells.

The most common solar cells are made from silicon, a semiconductor that is the second most abundant element on Earth. In a solar cell, crystalline silicon is sandwiched between conductive layers. Each silicon atom is connected to its neighbors by four strong bonds, which keep the electrons in place so no current can flow.

Here's the key: a silicon solar cell uses two different layers of silicon. An n-type silicon has extra electrons, and p-type silicon has extra spaces for electrons, called holes. Where the two types of silicon meet, electrons can wander across the p/n junction, leaving a positive charge on one side and creating negative charge on the other. You can think of light as the flow of tiny particles called photons, shooting out from the Sun. When one of these photons strikes the silicon cell with enough energy, it can knock an electron from its bond, leaving a hole. The negatively charged electron and location of the positively charged hole are now free to move around. But because of the electric field at the p/n junction, they'll only go one way. The electron is drawn to the n-side, while the hole is drawn to the p-side. The mobile electrons are collected by thin metal fingers at the top of the cell. From there, they flow through an external circuit, doing electrical work like powering a lightbulb, before returning through the conductive aluminum sheet on the back. Each silicon cell only puts out half a volt, but you can string them together in modules to get more power.

Twelve photovoltaic cells are enough to charge a cellphone, while it takes many modules to power an entire house. Electrons are the only moving parts in a solar cell, and they all go back where they came from. There's nothing to get worn out or used up, so solar cells can last for decades. So what's stopping us from being completely reliant on solar power? There are political factors at play, not to mention businesses that lobby to maintain the status quo.

But for now, let's focus on the physical and logistical challenges, and the most obvious of those is that solar energy is unevenly distributed across the planet. Some areas are sunnier than others. It's also inconsistent. Less solar energy is available on cloudy days or at night. So a total reliance would require efficient ways to get electricity from sunny spots to cloudy ones.

The efficiency of the cell itself is a challenge, too. If sunlight is reflected instead of absorbed, or if dislodged electrons fall back into a hole before going through the circuit, that photon's energy is lost.

The most efficient solar cell yet still only converts 46% of the available sunlight to electricity, and most commercial systems are currently 15 – 20% efficient. In spite of these limitations, it actually would be possible to power the entire world with today's solar technology. We'd need the funding to build the infrastructure and a good deal of space. Estimates range from tens to hundreds of thousands of square miles, which seems like a lot, but the Sahara Desert alone is over 3 million square miles in area.

Meanwhile, solar cells are getting better, cheaper, and are competing with electricity from the grid. And innovations, like floating solar farms, may change the landscape entirely. Thought experiments aside, there's the fact that over a billion people don't have access to a reliable electric grid, especially in developing countries, many of which are sunny.

So in places like that, solar energy is already much cheaper and safer than available alternatives, like kerosene. For say, Finland or Seattle, though, effective solar energy may still be a little way off.

Notes

（1）solar panel 太阳能电池板：用于将太阳能转化为电能的装置。

（2）n-type silicon n 型硅：具有额外电子的硅材料。

（3）p-type silicon p 型硅：具有额外电子空位（空穴）的硅材料。

（4）p/n junction 半导体中的正-负接面：n 型硅和 p 型硅交界处，形成了电势差，使得电子和空穴只能在一侧移动。

（5）external circuit 外部电路：连接太阳能电池板的电路，可以将电流引导至需要供电的设备中。

（6）photovoltaic cell 光伏电池：将光能直接转化为电能的装置，也就是太阳能电池。

（7）lobby 游说：指企业、团体向政府或其他机构施加影响力，以推动特定立场或利益。

（8）status quo 现状：指现有的状况或状态。在文中表示存在一些企业游说维持现有能源结构的情况。

（9）unevenly distributed 不均匀分布：指在地球表面不同地区，太阳能的接收情况存在差异，有些地方阳光充足，而有些地方则较少。

（10）dislodged electron 被移位的电子：指受到太阳能电池光照激发的电子，从原有位置移动的过程。

（11）floating solar farm 水上太阳能农场、浮动太阳能发电厂：创新的太阳能发电方式，将太阳能电池板安装在水体上以增加发电效率。

参考译文

第 6 课　太阳能电池板是如何工作的？

地球接收了很多太阳能：173 000 太瓦。这一数字是地球上人类所使用能源总和的一万倍。那么有没有可能有一天我们的世界会完全依赖太阳能？

为了回答这一问题，首先我们需要明白太阳能板是怎样将太阳能转化为电能的。太阳能板由更小的单位构成，称为太阳能电池。

最常见的太阳能电池原料是硅。硅是一种半导体，也是地球上储量第二的物质。在一个太阳能电池单元里，晶体硅像是三明治夹心一样被夹在两个导电层之间。每个硅原子通过四个强键和其他硅原子连接，能够固定电子，因此不会产生电流。

工作原理如下：一个硅太阳能电池有着两个不同的硅层。包括一层有着额外电子的 n 型硅层，和拥有能存储额外电子的空隙的 p 型硅层，称作"空穴"。在两种硅层交界处，电子可以在 p-n 的连接处移动，在一个硅层上产生正带有负电荷的电子和带有正电荷的空隙现在可以自由移动了。但是由于 p 型和 n 型硅层交界处存在电场，它们只会向一个方向移动。电子被吸引至 n 型硅层处，空穴则跑到 p 型硅层处。这些自由电子被太阳能电池顶部的细小金属导体收集。从那里开始，它们流过一个外部电路，以电流的形式做功，比如点亮一盏电灯，然后从背面的铝制导体薄片返回。每个硅太阳能电池只能产生半伏特的电压，但你可以将它们组装成模块来获得更多的能量。

十二个光电太阳能电池足以为手机充电，但是为整幢房屋提供电能需要许多太阳能板。电子是太阳能电池中唯一的移动单元，而且它们工作后会沿原路返回。没有组件会遭受磨损和损伤，因此太阳能电池可以使用几十年。那么是什么限制了我们完全依赖太阳能提供能源呢？这里有着政治因素在起作用，更不要说企业界为了维持利润现状而四处游说（反对使用太阳能）。

但现在，我们只研究物理和后勤方面的障碍，其中最主要的原因是太阳能不是均匀地分布在这颗星球上的。其中一些地区比其他地区更晴朗。而且阳光也不是持续性的。在多云天气或是晚上太阳能供给量就会下降。所以对太阳能的完全依赖需要从晴朗地区将电能转移到多云地区的高效方法。

太阳能电池本身的效率也是一大障碍。如果阳光被反射而不是被吸收，或是失去束缚的电子在进入电路前掉进了空穴中，这时光子的能量就损失掉了。

迄今为止最高效的太阳能电池也只能将阳光能量的 46% 转化为电能，而目前大多数商业化的太阳能系统只有 15—20% 的转化效率。即使面临着这些障碍，使用现如今的太阳能技术为全世界供电依然是可行的。我们只需要制造设施的资金和足够的空间。估计需要几万至几十万平方英里的土地，这看起来很多，但光是撒哈拉沙漠就有三百万平方英里的土地。

与此同时，太阳能电池正变得越来越高效、经济，而且正和输电网进行竞争。像水上太阳能农场之类的创新设计可能会完全改变当今的（能源）格局。这不是异想天开，事实上超过十亿民众还没有可靠的输电网供给，尤其是在发展中国家，其中许多国家阳光很充足。因此在这些地方，太阳能比起煤油等其他能源来说已经十分经济和安全了。

但至于芬兰和西雅图（这些阴雨连绵的地方），有效的太阳能应用可能还有很长的路要走。

英翻中练习

(1) The rapid development of renewable energy in China has made headlines around the world. In 2017, the installed capacity of solar PV power generation was 53 gigawatts, which is a 53.6% increase from 2016, ranking first in the world in terms of added solar power for five consecutive years. By the end of 2017, China's total PV solar capacity reached 130 gigawatts, ranking first in the world for the third year in a row. Let's take a look at the top three mega solar projects in China.

Longyangxia Dam Solar Park

Ranking third is Longyangxia Dam Solar Park. In 2013, the solar PV station was built with an installed capacity of 320 megawatts, covering nine square kilometers. In 2015, an additional 530 mega-watts was installed, covering a total area of fourteen square kilometers, making Longyangxia Dam Solar Park one of the world's largest solar power stations with a total installed capacity of 850 megawatts.

Located in western China's Qinghai Province, the solar power station is integrated with a local hydroelectric power station. The park is coupled to one of the hydroelectric turbines, which automatically regulates the output to balance the variable generation from solar power before dispatching power to the grid. This project not only helps solve problems of unstable solar power, but also helps conserve water and is a highly-efficient use of renewable energy.

Panda Solar Power Plant

Coming in at number two is the Top Runner Base Project affiliated with the Panda Solar Power Plant in the city of Datong in Shanxi Province. Viewed from above, the base has the shape of a giant panda, which is how it got its name. In June 2017, the first phase of the project was completed, adding 50 mega-watts of installed capacity to the power grid in Datong.

Tengger Desert Solar Park

And finally, the number one project is the Tengger Desert Solar Park in Zhongwei of Ningxia Hui Autonomous Region. It covers 1,200 square kilometers of land with an installed capacity of 1,547 mega-watts. The solar park is the world's largest solar array, known as the "Solar Great Wall" of China. Though this is currently the largest solar park in China, the Panda Power Plant in Datong will take the top spot once it is completed.

It is a shared mission of the international community to speed up the transition to renewable energy and achieve green, low-carbon development. China will continue to promote the development of green industries and seek a brighter and cleaner tomorrow for human beings.

(2) The West-to-East Electricity Transmission Project is a major part of the Chinese government's endeavor to develop the country's western region on a large scale as well as an important measure for balancing the development of eastern and western regions by leveraging their complementary advantages. Guizhou is a leading power supplier for the project's South Channel. As of December 2020, Guizhou had cumulatively transmitted more than 600 billion kilowatt-hours of electricity to Guangdong Province and its surrounding areas. In 2020, Guizhou supplied 50 billion kilowatt-hours of electricity to Guangdong, more than ten times that in 2002. Guizhou has become one of the industrial arteries of China, "Factory of the World", while powering the Pearl River Delta region.

In 2020, Guizhou's installed generating capacity reached 74.78 million kilowatts, with 232.7 billion kilowatt-hours of electricity generated, and clean energy accounted for more than half of the installed capacity. It is expected that the province's installed generating capacity will exceed 100 million kilowatts by 2025.

Text 7 Smart Grid Development in China

Evolution of the smart grid concept

In China, the concept and notion of what a smart grid is has constantly been evolving. Currently, the parties involved in smart grid research and development mainly consist of grid companies, academia, and government. Each of them has a different focus on certain aspects of a smart grid. Grid companies generally focus on safe grid development and grid stability in response to challenges arising from the significant addition of new electricity sources. Academia and social enterprises focus more on NE (New Energy), DG, and innovation in business models of grid users with the hope of promoting significant changes in grid development and operation. Government is mainly concerned with developing a secure, efficient, and clean modern energy system, with particular emphasis on promoting the integration of renewable energy and the development of DG. Furthermore, the implementation of the power market reform and a coordinated development of the large public power grid and microgrids are important for government.

1. Smart grid strategy of power grid companies

As the most important actor of China's power grid development and operation, State Grid Corporation of China (SGCC) occupies a pivotal position in smart grid research and development. SGCC mainly focuses on the development of ultra-high voltage (UHV) electrical grids from the perspective of grid security. Another important focus is the adoption of digital technology to improve dispatching and operation of the large public grids.

In 2014, SGCC took the lead in proposing the concept of "Global Energy Interconnection (GEI)", an upgrade of the original term "Strong Smart Grid". According to SGCC, GEI should be composed of an UHV grid, smart grids and clean energy sources. This highlights the role of a UHV grid as the backbone and a strong smart grid of global electrical interconnection as the core for promoting clean energy development and electrification.

Since 2016, and in response to the rapid development of renewable energy in China, the SGCC concept of a smart grid has gradually transitioned to the term "energy interconnection". Energy interconnection (or also Energy Internet) can be understood as the comprehensive use of advanced electronics, information, and intelligent management technologies to interconnect a large number of energy nodes. These nodes consist of distributed energy data monitoring devices, distributed energy storage devices and various types of loads, such as power, oil and natural gas networks. A two-way flow of energy through peer-to-peer exchange and energy sharing should be achieved. Energy interconnection underscores the application of modern information and intelligent control technology as well as Internet of Things (IoT) technology to realize the interconnection and interoperability of the power grid with the heat/gas/transportation networks. In that way the resilience and regulation capability of different energy systems should be enhanced.

2. The notion of smart grid in academia and private enterprises

In contrast to SGCC's efforts in achieving interconnection at higher voltage levels and maintaining a unified large grid pattern, academia places more emphasis on enabling extensive interaction between the distribution and the demand side. The promotion of DG is a further key research area. In April 2015, a symposium on "Energy Interconnection: Frontier Scientific Issues and Key Technologies", initiated by Tsinghua University, was held in Beijing. Over 50 prestigious scholars from 25 national and international institutions participated in the symposium. Their common view underlined the importance of the demand-side and DG in energy interconnection and grid development. The academic notion of smart grid thus focuses more on complementarity and fair peer-to-peer participation on the consumer side. The introduction of concepts from network and internet theory into traditional energy networks is reflected in the following aspects.

Open interconnection: open interconnection of a variety of energy sources; open, peer-to-peer access of various devices and systems; open participation of different actors and end users; open energy market and trading platforms; open data and standards.

User-centeredness: innovation in business models centered on users' personalized and diversified needs; user recognition and extensive participation to effectively enhance the value of energy production, operation, trading, and services; user-centeredness is reflected mainly in user experience.

Distributed energy: distributed energy as important driving force for energy interconnection development; development of DG turns an increasing number of grid users into prosumers (both consumers and producers); demand for DG plug-and-play solutions and user needs for microgrid development will break the monopoly of grid operation; reciprocity of energy production and consumption as well as the formation of network-like transaction relationships.

3. Government position and support for smart grid

The Chinese government understands smart grid development as an important part of upgrading the power grid. Crucial tasks of smart grids in that respect are smart dispatching, upgrading distribution grid digitalization and the integration of DG and NE. Government has a holistic approach considering the development requirements of both large public and microgrids.

Since 2015, the government and relevant ministries, such as National Development and Reform Commission (NDRC), National Energy Administration (NEA) and MIIT (Ministry of Industry and Information Technology) published guiding opinions and plans on smart grid developments.

Smart Grid Progress and Achievements

Since 2014, China has stepped up efforts in its smart grid development. Although China's electricity system is still dominated by large public grids with distinctive features of a planned economy, new factors influencing smart grids have significantly increased. The factors are manifested in the strong development of renewable energies, the emergence of distributed generation (DG) and the increased demand for a fair market environment on the consumer side.

At the end of 2021, the total installed renewable energy capacity exceeded 1,000 gigawatt (GW). The strong development of modern renewable energies, represented by wind and solar photovoltaic (PV) power, is the most significant feature of China's smart grid. The country has the highest wind and solar capacity globally. As China proposed to achieve carbon neutrality by 2060, the large-scale expansion of new energy sources, mainly wind and solar as variable renewable energy (VRE), has become an essential pathway for achieving this ambition. However, since VRE is characterized by fluctuations in generation, the challenges of grid regulation and balancing electricity production and consumption is the first to be addressed by the smart grid.

Renewable energies are distributed over wider geo-graphical areas. Thus, the need arises for transmitting and distributing the generated electricity over longer distances. Developing DG in the vicinity of load centers often becomes the most economical way of renewable energy development and is an increasingly important option. While the development of DG enhances demand-side capabilities to participate in grid regulation and power market transactions, it also complicates the traditional power dispatching and operation mechanism. This impacts the traditional power system governance and raises questions on how to reconcile the interests of different actors.

Despite the smart grid technology developments and market reforms being underway, China still faces challenges:

Expanding NE comes with difficulties in grid balancing and secure operation;

Operational and implementation barriers to establish market mechanisms supporting the development of VRE and DG;

Digital applications and utilization are uncommon, reflecting the early stages of a digital energy sector.

Outlook for Smart Grid Development in China

(1) Development of different energy storage technologies that enhance the capability for VRE integration. Currently the most cost-effective energy storage method

is pumped-hydro power. China has launched ambitious targets for pumped-hydro power storage development. The advancement and cost reduction of new energy storage technologies such as batteries can enable new energy storage technologies to become a key force in energy storage. The vast number of electric vehicles (EV) can serve as important demand-side energy storage sources and flexibility given appropriate guiding mechanisms.

(2) Support the development of flexible power grids through local, DG-centered balancing. The development of DG will contribute to a more diversified pattern of China's power grid. It is expected that by 2030, China's installed capacity of distributed renewables will reach 400 GW. This growth can be accommodated with a large amount of local microgrids operating in coordination with the public transmission grids. The internal micro-balance of the microgrids will be important for achieving the balance of the entire power system.

(3) Extensive application of digital technologies in all areas of the electricity system. Digital technologies can support a more efficient smart grid operation. They promote management improvements and business transformation within the power sector. The main applications are online equipment monitoring and control; smart operation and maintenance; comprehensive transaction support; planning, simulation, and operation of power systems with a high proportion of NE.

(4) Creation of a wide variety of new demand- and user-side entities to allow for diversified market participation modes. Examples are the interaction between EVs and the grid or the participation of aggregated controllable loads through virtual power plants (VPPs) in grid regulation.

(5) Development of a sound green electricity trading system. Compared with traditional energy sources, renewable energy has the advantages of being low carbon and environmentally friendly. However, renewable energy also has disadvantages (e. g. fluctuating generation) that make it significantly less competitive in view of the entire power system. To make up for this lack of systemic competitiveness, a fully-fledged green power trading system for renewable energy should gradually be put in place.

To accelerate the development of smart grids, the support for novel energy storage, digital solutions, and smart microgrid technologies needs to be stepped up. Market mechanisms and their design need to accommodate a high proportion of NE in the new power system. Policies for supporting green electricity, green certificates trading, and carbon markets need to be improved to highlight the environmental value of renewable energy and incentivize enterprises to actively purchase green electricity.

Notes

(1) VRE (variable renewable energy)可变可再生能源：指波动性较高的可再生能源，如风能和太阳能。

（2）dispatch 调度：在电力系统中，指根据需求和供给条件，合理分配和控制电力输出的过程。

（3）microgrids 微电网：指小型的能源系统，可以与大型公共电网连接，也可以独立运行，通常由分布式发电、能量储存和能源管理系统组成。

（4）Global Energy Interconnection（GEI）全球能源互联网：中国国家电网公司在2014年首次提出的概念，是对原始术语"强智能电网"的升级。GEI应由特高压电网、智能电网和清洁能源组成，强调特高压电网作为智能电网和全球电力互联的骨干，促进清洁能源发展和电气化。

（5）peer-to-peer exchange 点对点交流：指能源节点之间直接进行能源交流和共享，实现能源的直接传输和共享，而不经过中央调度。

（6）Internet of Things（IoT）物联网：指通过互联网将各种物理设备、传感器、软件等连接起来，实现设备之间的信息互通和智能化管理。在能源互联中，物联网技术用于实现能源节点的互联互通和智能控制。

（7）open interconnection 开放互连：指各种能源来源的开放互联，各种设备和系统开放的点对点访问，不同参与者和终端用户的开放参与，开放能源市场和交易平台，开放数据和标准。

（8）digital energy sector 数字能源部门：指利用数字技术和信息通信技术进行能源管理和运营的领域，包括智能电网、电力市场、能源大数据等。

（9）aggregated controllable loads 聚合可控负荷：将多个可控负荷（如家电、工业设备等）进行集中管理和控制，以在电网调节中提供灵活性。

（10）green power trading system 绿色交易体系、绿色电力交易体系：是为可再生能源建立的完善的电力交易机制，以促进其市场竞争力。

参考译文

第7课　中国智能电网的发展情况

智能电网内涵的变迁

在中国，关于智能电网的概念，一直处于调整变化过程中。当前，参与智能电网研究和建设的主体主要包含：电网企业、学界和政府等。这些主体对智能电网关注的侧重点各不相同。电网企业关注大电网的安全发展，以应对新能源电力显著增长带来的挑战；学术界与社会企业更加侧重新能源、分布式电源以及用户侧业务模式创新，以促进电网发展的显著变革；政府侧重安全、高效、清洁的现代能源保障体系建设，重点促进可再生能源消纳、分布式能源发展、电力市场改革等方面，注重大电网与微电网的协同发展。

1. 电网企业的智能电网战略

作为中国电网建设运行最重要的成员，国家电网在智能电网的研究和建设方面占据举足轻重的地位，国家电网主要从保电网安全角度出发，注重特高压电网的建设，以及采用数字化技术提升大电网的调度运行水平。

2014年，国家电网公司首次提出"全球能源互联网"的概念，该概念是在原来"坚强智能电网"基础上的提升。根据国家电网的阐述，全球能源互联网是由智能电网、特高压电网和清洁能源三方面内容构成,强调建立以特高压电网为骨干网架、全球互联的坚强智能电网作为推动清洁能源发展和电能替代的核心。

2016年起,为应对可再生能源的快速发展,国家电网对智能电网的概念逐步过渡到"能源互联网"的表述,以增强现代信息技术、物联网技术、智能控制技术等新技术的应用,实现电网与热网、燃气网、交通网的互联、互通,提升能源系统的韧性与调节能力。

2. 学术界与社会企业对智能电网的理解

相对国家电网公司强调实现更高电压等级互联,维持统一大电网格局的目标,学术界更重视在配网侧和用户侧的广泛互动以及促进分布式能源的发展。2015年4月,由清华大学发起的香山科学会议"能源互联网:前沿科学问题与关键技术"学术讨论会在北京召开,来自25家单位的50多位海内外知名学者参加了会议,形成了以用户侧、分布式电源为主的能源互联网发展观点。学术界对智能电网的定义更侧重于多能互补、用户侧的公平对等参与,以及在传统能源网基础上引入互联网理念,主要内涵如下:

开放互联:多种能源的开放互联,各种设备与系统的开放对等接入,各种参与者与终端用户的开放参与,开放的能源市场和交易平台,开放的数据和标准等。

以用户为中心:以用户个性化、多元化需求为中心进行商业模式创新。用户认可和广泛参与能有效增强能源生产、运行、交易、服务等各环节的价值。以用户为中心强调用户体验。

分布式能源:分布式能源是推动能源互联网发展的重要动力,分布式电源的发展使更多用户成为产销者(既是消费者,又是生产者),用户对于分布式电源即插即用的主张,以及发展智能微电网的需求,将打破电网经营的垄断,实现能源生产、消费的对等以及网状交易关系的产生。

3. 政府对智能电网的定位与支持

中国政府将智能电网建设作为电网升级的一项重要内容,将电网智能调度、配电网智能化升级、分布式能源与可再生能源消纳作为智能电网的重要任务,兼顾了大电网与微电网的发展要求。

2015年7月,国家发改委、能源局发布《关于促进智能电网发展的指导意见》,定义智能电网是在传统电力系统基础上,通过集成新能源、新材料、新设备和先进传感技术、信息技术、控制技术、储能技术等新技术,形成的新一代电力系统,发展智能电网是发展能源互联网的重要基础。文件提出了提升电网智能化水平、提高新能源消纳能力、引导用户参与节能减排等10项任务。

中国智能电网的进展与成就

2014年以来,中国智能电网发展加快,尽管中国电网仍以公用大电网为主导、计划经济特征依然显著,但智能电网新的影响因素明显加强,主要反映在可再生能源的跨越式发展、分布式电源的涌现以及用户侧对公平市场环境的要求提高。主要体现在:

首先,以风电、光伏为代表的可再生能源的跨越式发展是中国智能电网最显著特征。截至2021年底,中国可再生能源装机规模突破10亿千瓦,其中风电、太阳能装机均跃升至世界第一。中国提出2060年前实现碳中和的目标,风电、太阳能为主的可变可再生能源的规模化发展是实现碳中和的重要途径。但可变可再生能源的功率具有随机波动性的特点,如何解决电网调峰问题和发用电平衡问题,是发展智能电网的首要挑战。

其次，可再生能源具有分布广、密度低的特点，在靠近负荷的地方建设分布式电源成为可再生能源发展的最经济的方式，分布式电源的地位日益重要。分布式电源的发展，增强了用户侧参与电网调节和电力市场交易的能力和意愿，但也增加了传统电力调度运行机制的负担，也对传统的电力系统利益格局带来了冲击。

尽管中国智能电网在技术发展和市场化改革方面都取得了突破性的进展，但仍存在来自技术、机制等方面较多的挑战。首先是新能源的发展带来电网平衡困难和安全挑战；其次，建立支撑可再生能源、分布式能源发展的市场机制仍面临诸多束缚；再次，数字化的应用和推广仍不普遍，数字化业务水平仍处于初级阶段。

中国智能电网的发展方向

第一，发展多种储能技术，提升可变可再生能源的消纳能力。当前最经济有效的储能是抽水蓄能，中国推出了颇具雄心的抽水蓄能发展目标。但未来随着电化学等新型储能技术的进步和成本的下降，新型储能也将成为储能的主力。同时，用户侧海量电动汽车，在合适的引导机制下也将成为重要的储能资源。

第二，以分布式电源为核心的局域微平衡支撑电网柔性发展。分布式电源的发展，将给中国电网带来更多元的发展趋势。预计至2030年，中国分布式新能源装机容量将达到4亿千瓦，将形成海量的微电网与公用电网协同运行的格局，微电网内部的微平衡成为整个电力系统平衡的重要基础。

第三，数字化技术全方面应用于电力生产，全面支撑智能电网业务更高效运营，促进电力企业管理提升和业务转型。其主要应用包括：设备的在线监测、控制；智能运维的开展；对交易的全面支撑；高比例新能源电力系统的规划、仿真与运行等。

第四，用户侧新型主体广泛发展，形成多元的市场参与方式。包括电动汽车与电网的互动，通过虚拟电厂聚合可控负荷参与电网调节等。

第五，绿色交易体系发展健全。与传统能源相比，可再生能源具有低碳、绿色环保优势，但可再生能源调节性能差，在电力市场竞争中将处于明显的劣势。为了弥补可再生能源竞争力的不足，将逐步健全可再生能源的绿色交易体系。

为了促进智能电网的更快发展，建议加大新型储能、数字技术、智能微电网等技术的支持力度。同时，在市场机制上，健全适应高比例新能源新型电力系统的市场机制，完善绿电-绿证-碳市场协同政策，凸显可再生能源的环境价值，促进企业积极采购绿电。

英翻中练习

（1）Around 2008, China put forward a smart grid development target that was geared towards increased automatization levels in terms of power generation, transmission, distribution, and consumption. Compared with European countries and the U.S., China's power grid has the following characteristics: The main goal of China's power grid development is and will be, for a considerable length of time, to meet the fast-growing electricity demand, to continue to strengthen the grid structure dominated by large power networks and to improve the power supply system with certain aspects of a planned system. Since 2014, China's electricity demand has been growing at a fast pace, and grid operators still operate in the inertia of the old large public grid development model. Smart grid development, however, increasingly gained traction due to new influencing factors: the strong development of renewable energies, the emergence of distributed generation (DG), and the increased demand for a fair market environment on the demand-side. These factors require an improved level of digitalization and market-mechanisms for the grid and the power system. As significant changes take place in the structure and shape of China's power system, smart grid developments are increasingly shaped by a coordinated development of centralized and distributed power sources.

China has also long been committed to implementing a power market reform, using price signals to promote active participation of all actors, breaking the traditional large grid-based governance pattern, releasing the power industry's innovative potential, and providing institutional safeguards for smart grid development. This progressive power market reform has become an important prerequisite for the development of smart grids, especially regarding the establishment of a unified power market adapted to the increase of renewable energy. The formation of a fair market environment with broad participation from all grid users is crucial for enabling a transition towards a green, smart, and open grid.

(2) VPPs are important actors for an effective management of DG connections to the grid. Their unique business model of aggregating DG, controllable loads, and energy storage devices at the distribution level, allows VPPs to participate in grid operation and dispatching. VPPs are relevant in mediating the conflicts between grid operation and DG by fully reflecting the value and benefits DG brings to the grid and users.

With the development of VPPs, loads also become essential components of smart grids. VPPs can exploit flexibility potentials at the consumers and their loads. A dynamic regulation of VRE and peak load shaving can ultimately lead to reduced investment needs into power generation and the grid. According to Chinese government documents, future demand response capacity needs to reach 5%-8% of the maximum load. In 2030, the demand response capacity is anticipated to exceed 100 GW.

Text 8 Issues and Challenges of Energy Engineering

It's easy to think, from the Western perspective, that the great days of engineering were in the past during the era of massive mechanization and urbanization that had its heyday in the nineteenth century and which took the early Industrial Revolution from the eighteenth century right through into the twentieth century which, incidentally, simultaneously improved the health and well-being of the common person with improvements in water supply and sanitation. That era of great engineering enjoyed two advantages: seemingly unlimited sources of power, coal, oil and gas, and a world environment of apparently boundless capacity in terms of water supply, materials, and other resources relative to human need.

Now we know differently. We face two issues of truly global proportions—climate change and poverty reduction. The tasks confronting engineers of the twenty-first century are:

Engineering the world to avert an environmental crisis caused in part by earlier generations in terms of energy use, greenhouse gas emissions, and their contribution to climate change, and Engineering the large proportion of the world's increasing population out of poverty, and the associated problems encapsulated by the UN Millennium Development Goals.

This will require a combination of re-engineering existing infrastructure together with the provision of first-time infrastructure at a global scale.

And the difference between now and the nineteenth century? This time the scale of the problem is at a greater order of magnitude; environmental constraints are dangerously close to being breached; worldwide competition for scarce resources could create international tensions; and the freedom to power our way into the future by burning fossil fuels is denied.

Resolving these issues will require tremendous innovation and ingenuity by engineers, working alongside other technical and non-technical disciplines. It requires the engineer's ability to synthesize solutions and not simply their ability to analyze problems. It needs the engineers' ability to take a systems view at a range of scales, from devices and products through to the large-scale delivery of infrastructure services.

This means that the great age of engineering is NOW.

The world will have to bring about a better quality of life to substantial numbers of people by providing access to affordable energy and, at the same time, it will need to mitigate and adapt to climate change. Effective and feasible choices of energy source, energy technology, and end-user efficiency will make a major contribution to addressing this challenge. These choices will have to be made in a transparent and professional manner. Several branches of engineering will play key roles even in the assessment and decision phases of this process.

Two billion people are without access to affordable and clean energy services and as many again are without reliable access. Yet, access to energy is key to achieving all of the Millennium Development Goals. To meet basic human needs, it will be necessary to provide energy for all through access to reliable and affordable energy services, giving particular attention to the urban and rural poor.

Climate change is recognized as a global sustainable development challenge with strong social, economic, and environmental dimensions. Climate change is attributed to anthropogenic sources—excessive greenhouse gas emissions from human energy production and consumption.

It is widely acknowledged that existing solutions are not yet sufficient for meeting the world's growing energy needs in a sustainable manner. Though energy technologies are rapidly developing, much work and innovation are still needed to bring about substantial changes in energy for heating, transportation, and electricity as well as in energy efficiency, conservation, and behavior. It is essential to change unsustainable patterns of consumption, and this will require difficult cultural adjustment in some countries.

Determining the technological, economic, and environmental feasibility of an energy option is one of the chief roles asked of the engineering profession. Engineers are actively involved in the development and implementation of technologies used to generate energy; indeed, engineers design, build, operate, maintain, and decommission the energy systems of the world. Sustainable energy policies need to conform to realistic and factual conditions; scientifically sound and thoroughly engineered solutions are the only way to address the issues of energy sustainability.

To meet basic human needs and facilitate the achievement of the Millennium Development Goals, it will be necessary to provide energy for all and access to reliable and affordable energy services, giving particular attention to the rural and urban poor.

Energy is crucial for sustainable development. A sophisticated energy mix that employs mature and feasible technologies will be needed in most countries. Ambitious but acceptable limits for greenhouse gas emissions must be managed, calling for ever-greater international cooperation. Priority should be given to exploring carbon sequestration schemes for fossil fuel utilization, ensuring the highest state-of-the-art standards of safety and non-proliferation for nuclear energy, innovation for higher efficiencies of renewable energies, designing compromises for agricultural land usage and population displacement for hydropower, and developing technologies for energy efficiency and conservation. Decisions on the use of a given technology that could contribute to sustainable energy development require a thorough analysis of technological and economic feasibility; the technology of a proposed solution should be available at the time the need for it becomes apparent, and the energy it provides should be affordable for the majority of the population.

There exists a relationship between quality of life and per capita consumption of energy. Indeed, the Human Development Index shows a close link between increasing quality of life in a given country and an increase in energy use per capita. In general, a high quality of life is currently achieved with a per capita consumption of 100 billion to 150 billion joules of energy. If the countries currently exceeding this level could decrease their energy consumption to within this range, their quality of life would be maintained, and global resources would be better preserved and utilized, particularly among countries at different stages of development.

Significant differences exist between developed and developing countries, and energy policies must be very context-specific; they do not translate from one country to another. There is no universal solution for making sustainable energy available globally, but developing countries can learn from the lessons of developed countries. The optimal energy mix for any country will depend on, among others, its available natural resource base, population distribution, predicted growth of energy demand, and its engineering and economic capacity. Energy solutions for developed countries are not always adequate in developing countries because, for example, developing countries often see much higher annual growth rates in demand.

In developed countries, growth in demand is more stable, at around 1 per cent to 2 per cent per year, than in developing countries where it can reach about 4 per cent to 5 per cent. Add to this the phenomenon of strong urban migration, and huge energy needs are being further concentrated in emerging mega-cities. Electricity grids in some developed countries have experienced major national and international failures due to a lack of capacity and investment; even "stable" rates of energy demand can cause instabilities in energy supply.

Predictions of energy consumption in developed and developing countries show that, in a short space of time, demand for primary energy in developing countries will overtake that of developed countries. Sooner or later this may cause supply disturbances to developed countries, since much of the energy they are exploiting in developing countries will be taken out of the export market to satisfy local demand.

Biofuels (separate from biogas, biomass, and so on) have been developing in many countries, usually as fuel additives, to increase energy security, reduce greenhouse gas emissions, and stimulate rural development. However, economic, social, and environmental issues limit the extent to which these goals can be met with current biofuel technologies; there are serious concerns about whether they do, in fact, reduce greenhouse gases overall as well as the effect they are having on land use, biodiversity, and food prices.

First-generation biofuels, such as bioethanol and biodiesel, are only economically competitive with fossil fuels in the most efficient agricultural production markets and under favorable market conditions of high oil prices and low feedstock prices. One of the potential new risks for drylands is growing biofuel crops using unsustainable cultivation practices, leading to accelerated soil erosion and desertification. Growing biofuel crops-using sustainable cultivation practices-on semi-arid and sub-humid lands unsuitable for food production would not compete with food production and could help rehabilitate those lands.

A shift towards cellulose-based second-generation biofuels, using wood and grassy crops, would offer greater net reductions in emissions and use less land, but technical breakthroughs are required. The potential for second-generation biofuels that are economically, environmentally, and socially sustainable needs to be thoroughly researched, involving modern agricultural engineering tools.

The main challenges the world is going to face in the near future are centered on the explosion of energy demand, mainly in developing countries, and on the constraints imposed by climate change on greenhouse gas emissions that will need to be drastically abated.

Given that fossil fuels will continue to play a dominant role in the energy mix in the decades to come, the development and use of advanced and cleaner fossil fuel technologies should be increased. Hybrid technologies that use both fossil fuels and other energy sources may become more affordable and feasible on a larger scale.

There is also considerable scope for improving energy efficiency in households, transport, and industry. Energy efficiency and economy are fundamental to reducing greenhouse gas emissions and are increasingly important for the financial economy with the cost of energy rising around the world. Many energy efficient technologies, however, require the use of more complex and uncommon systems and materials and have long pay-back times; these are important considerations before they can be viable for the population at large.

At present, the cost of energy from renewable and sustainable sources is higher than the cost of energy from non-renewable and unsustainable sources (although this is the subject of debate). Renewable sources (such as solar photovoltaic, wind, and hydropower), even without the benefit of economies of scale, are ideal for use in situations where the energy demand is growing slowly or in places far from high-consumption centers. Engineering efforts are currently needed to lower the cost of renewable generation and also to find feasible technologies for renewable energy storage (such as hydrogen fuel cells) and distribution (such as distributed generation for small-scale renewable sources).

In the transportation sector, actions for promoting cleaner fuels and vehicles must be complemented by policies to reduce the overall demand for personal vehicle use, particularly by encouraging public transport. Modifying unsustainable transportation energy consumption patterns will require politically difficult cultural adjustments.

There is great urgency to design and implement measures for both mitigation and adaptation towards unavoidable climate change effects, including upgrading infrastructure to withstand the impacts of extreme weather events and "climate proofing" of new projects. Development, deployment, and diffusion of low-carbon energy technologies, together with energy efficiency, renewable energy, and cleaner and advanced technologies for energy supply, will require intense engineering ingenuity.

Notes

(1) energy mix 能源结构：指一个特定地区或国家所使用的各种能源类型和比例。

(2) economies of scale 规模经济：指在产量增加的情况下，单位成本随着规模扩大而减少的经济现象。

（3）distributed generation 分布式发电：指将能源发电设备分散布置在用户端或接近用户的地方。与集中式发电相比，分布式发电可以提高能源的可再生利用率，减少能源传输损耗并增强能源系统的韧性。

（4）climate proofing 气候适应、防灾气候设计

参考译文

第8课　能源工程的问题与挑战

从西方的角度来看，人类工程学的辉煌时代似乎已经过去了，在十九世纪达到鼎盛的大规模机械化和城市化时代，使早期工业革命从十八世纪延续到了二十世纪，这同时也附带改善了普通人的健康和福祉，改善了供水和卫生设施。那个伟大的工程学时代享有两大优势：看似无限的能源来源（煤炭、石油和天然气），以及相对于人类需求而言，水源、材料和其他资源似乎具有无限的容量。

现在我们的看法不一样了。我们面临两个真正全球性的问题——气候变化和减贫。21世纪工程师面临的任务有两个方面：

一方面，通过改变工程方法来避免环境危机，这部分危机部分源自早期世代的能源使用、温室气体排放量以及对气候变化的影响；

另一方面，减少全球越来越多的贫困人口，并解决联合国千年发展目标所涵盖的相关问题。

为实现这一目标，需要在全球范围内改建现有基础设施，同时提供首次基础设施。

那么现在和19世纪的区别是什么呢？这次问题的规模更大；环境制约因素即将被危险地突破；全球范围内对有限资源的竞争可能引发国际紧张局势；并且我们被剥夺了利用化石燃料来推动未来发展的自由。

解决这些问题需要工程师的巨大创新和聪明才智，与其他技术和非技术学科一起合作。这需要工程师能具有整合解决方案的能力，而不仅仅是分析问题的能力。需要工程师具备在各个层面上考虑问题的系统观，从设备和产品到基础设施服务的大规模提供。

这意味着工程学的伟大时代就是现在。

世界需要通过提供可负担得起的能源来改善大量人民的生活质量，并同时减缓并适应气候变化。有效而可行的能源、能源技术和终端用户效率的选择将对应对这一挑战起到重要作用。这些选择必须以透明和专业的方式做出。在此过程中，几个工程分支甚至在评估和决策阶段都将发挥重要作用。

20亿人口无法获得可负担和清洁的能源服务，同样多的人也无法获得可靠的能源供应。然而，能源的获取对于实现所有千年发展目标至关重要。为了满足人类基本需求，需要通过提供可靠和负担得起的能源服务为所有人提供能源，要特别关注城市和农村贫困人口。

气候变化被认为是一个具有重要社会、经济和环境维度的全球可持续发展挑战。气候变化归因于人为因素，即人类能源生产和消费过程中过量的温室气体排放。

广泛认可的是，现有的解决方案对于以可持续方式满足世界日益增长的能源需求还不够。虽然能源技术正在快速发展，但仍然需要大量的工作和创新，以在供暖、交通和电力方面实现实质性的变革，同时在能源效率、节约和行为方面做出改变。改变不可持续的消费模

式是至关重要的,这将需要一些国家进行艰难的文化调整。

确定能源选择的技术、经济和环境可行性是工程专业的主要职责之一。工程师积极参与到用于发电的技术的开发和实施中;事实上,工程师设计、建造、操作、维护和除役全球能源系统。可持续能源政策需要符合现实和事实的条件;科学上可行和经过周密工程设计的解决方案是应对能源可持续性问题的唯一途径。

为满足基本人类需求、推动实现千年发展目标,提供全民能源以及可靠、可负担的能源服务至关重要,特别要关注农村和城市贫困人口。

能源对可持续发展至关重要。大多数国家需要采用成熟而可行的技术实现复杂的能源结构。需要更大范围的国际合作来管理雄心勃勃但可接受的温室气体排放限额。优先考虑探索碳封存方案以利用化石燃料,确保核能安全和非扩散最高现代化标准,创新提高可再生能源的效率,就农业用地利用和水电造成的人口迁移进行折中,以及开发能源高效和节约的技术。对于可促进可持续能源发展的技术的使用决策,需要进行技术和经济可行性的彻底分析;在需求变得明显时,所提议的解决方案的技术应该是可获得的,并且其提供的能源应对大多数人口来说是负担得起的。

人类生活质量与人均能源消费之间存在着关联。事实上,人类发展指数显示了一个国家的生活质量的提高和人均能源使用增加之间的密切联系。一般而言,高生活质量通常是在人均能源消费为每年1000亿至1500亿焦耳之间达到的。如果超过这个范围的国家能够将能源消费降低至该范围内,它们的生活质量将得到保持,并且全球资源将得到更好的保护和利用,特别是对于那些处于不同发展阶段的国家更是如此。

发达国家和发展中国家存在着显著差异,能源政策必须完全因国而异,不能从一个国家转化到另一个国家。在使可持续能源在全球范围内可得的问题上,没有通用解决方案,但发展中国家可以借鉴发达国家的经验教训。任何国家的最佳能源结构将取决于其可利用的自然资源基础、人口分布、预计的能源需求增长以及其工程和经济能力等。发达国家的能源解决方案在发展中国家通常不够适用,因为发展中国家的能源需求往往具有高得多的年增长率。

在发达国家,能源需求的增长更加稳定,大约每年增长1%至2%,而在发展中国家,能源需求增长率可以达到约4%至5%。加上强烈的城市迁移现象,巨大的能源需求更进一步集中在新兴的大城市。一些发达国家的电力网络由于发电量不足和缺乏投资而经历了重大的国内和国际故障;即使是"稳定"的能源需求增长也可能导致能源供应的不稳定。

对发达国家和发展中国家能源消耗的预测显示,在短期内,发展中国家的一次能源需求将超过发达国家。这可能迟早会干扰发达国家的能源供应,因为发展中国家开采的大部分能源将被从出口市场拿出以满足本地需求。

生物燃料(与沼气、生物质等不同)已在许多国家开发,通常作为燃料添加剂,以增加能源安全性、减少温室气体排放和促进农村发展。然而,经济、社会和环境问题限制了目前生物燃料技术实现这些目标的程度;人们对其是否真正减少温室气体总体排放以及其对土地利用、生物多样性和食品价格的影响存在严重担忧。

第一代生物燃料,如生物乙醇和生物柴油,在最高效的农业生产市场和高油价、低原料价格的有利市场条件下,才能与化石燃料有经济竞争力。对于干旱地区而言,一个潜在的新风险是使用不可持续的耕作方式种植生物燃料作物,导致土壤加速侵蚀和沙漠化。在半干旱和亚湿润的不适宜粮食生产的土地上,种植生物燃料作物(采用可持续的耕作方式)将不会与粮食生产竞争,并有助于恢复这些土地。

利用木材和草本作物,转向以纤维素为基础的第二代生物燃料,可以更大程度地减少净排放并使用较少的土地,但需要技术突破。对于经济、环境和社会可持续性的第二代生物燃料潜力需要进行彻底研究,涉及现代化农业工程工具。

近期世界面临的主要挑战集中在能源需求的爆炸式增长,尤其是在发展中国家,以及气候变化对需要大幅度减少的温室气体排放的限制上。

考虑到化石燃料在未来几十年仍将在能源结构中发挥主导作用,应该增加先进且更清洁的化石燃料技术的开发和使用。同时使用化石燃料和其他能源的混合技术可能会在更大范围上变得更具经济性和可行性。

在家庭、交通和工业领域,提高能源效率有很大的潜力。能源效率和能源经济对减少温室气体排放至关重要,并且因全球能源成本上涨对金融经济而言愈发重要。然而,许多高效能源技术需要使用更复杂和不常见的系统和材料,并具有较长的投资回收时间;这些都是在大众中推行之前的重要考虑因素。

目前,可再生和可持续能源的成本比非可再生和不可持续能源更高(尽管这是有争议的)。在需求增长缓慢或远离高能耗中心的地方使用可再生能源(如太阳能光伏、风能和水能)即使没有规模经济的效益也非常理想。目前需要工程上的努力降低可再生能源发电的成本,并找到可行的可再生能源储存技术(如氢燃料电池)和配电技术(如小规模可再生能源的分布式发电)。

在交通领域,促进清洁燃料和车辆的行动必须与减少个人车辆使用的整体需求的政策相结合,特别是鼓励使用公共交通工具。改变不可持续的交通能源消费模式将需要政治上困难的文化调整。

迫切需要设计和实施应对无法避免的气候变化影响的减缓和适应措施,包括升级基础设施以抵御极端天气事件的影响,并对新项目进行"气候适应"。发展、应用和推广低碳能源技术,以及能源效率、可再生能源和清洁和先进能源供应技术,将需要极大的工程智慧。

英翻中练习

(1) World's first comprehensive energy roadmap shows government actions to rapidly boost clean energy and reduce fossil fuel use can create millions of jobs, lift economic growth and keep net zero in reach.

The world has a viable pathway to building a global energy sector with net-zero emissions in 2050, but it is narrow and requires an unprecedented transformation of how energy is produced, transported and used globally, the International Energy Agency said in a landmark special report released today.

Climate pledges by governments to date—even if fully achieved—would fall well short of what is required to bring global energy-related carbon dioxide (CO_2) emissions to net zero by 2050 and give the world an even chance of limiting the global temperature rise to 1.5℃, according to the new report, *Net Zero by 2050: A Roadmap for the Global Energy Sector*.

The report is the world's first comprehensive study of how to transition to a net zero energy system by 2050 while ensuring stable and affordable energy supplies, providing universal energy access, and enabling robust economic growth. It sets out a cost-effective and economically productive pathway, resulting in a clean, dynamic and resilient energy economy dominated by renewables like solar and wind instead of fossil fuels. The report also examines key uncertainties, such as the roles of bioenergy, carbon capture and behavioural changes in reaching net zero.

(2) "Our Roadmap shows the priority actions that are needed today to ensure the opportunity of net-zero emissions by 2050—narrow but still achievable—is not lost. The scale and speed of the efforts demanded by this critical and formidable goal—our best chance of tackling climate change and limiting global warming to 1.5℃—make this perhaps the greatest challenge humankind has ever faced," said FatihBirol, the IEA Executive Director. "The IEA's pathway to this brighter future brings a historic surge in clean energy investment that creates millions of new jobs and lifts global economic growth. Moving the world onto that pathway requires strong and credible policy actions from governments, underpinned by much greater international cooperation."

Text 9 Projects and Project Management

Projects are one of the principal means by which we change our world. Whether the goal is to split the atom, tunnel under the English Channel, introduce Windows 10, or plan the Olympic Games, the means through which to achieve these challenges remains the same: project management. Project management has become one of the most popular tools for organizations, both public and private, to improve internal operations, respond rapidly to external opportunities, achieve technological breakthroughs, streamline new product development, and more robustly manage the challenges arising from the business environment. Consider what Tom Peters, best-selling author and management consultant, has to say about project management and its place in business: "Projects, rather than repetitive tasks, are now the basis for most value-added in business." Project management has become a critical component of successful business operations in worldwide organizations.

One of the key features of modern business is the nature of the opportunities and threats posed by external events. As never before, companies face international competition and the need to rapidly pursue commercial opportunities. They must modify and introduce products constantly, respond to customers as fast as possible, and maintain competitive cost and operating levels. Does performing all these tasks seem impossible? At one time, it was. Conventional wisdom held that a company could compete using a low-cost strategy or as a product innovator or with a focus on customer service. In short, companies had to pick their competitive niches and concede others their claim to market share. In the past 20 years, however, everything turned upside down. Companies such as General Electric, Apple, Ericsson, Boeing, and Oracle became increasingly effective at realizing all of these goals rather than settling for just one. These companies seemed to be successful in every aspect of the competitive model: They were fast to market and efficient, cost-conscious, and customer-focused. How were they performing the impossible?

Obviously, there is no one answer to this complex question. There is no doubt, however, that these companies shared at least one characteristic: They had developed and committed themselves to project management as a competitive tool. Old middle managers, reported Fortune magazine, are dinosaurs, and a new class of manager mammal is evolving to fill the niche they once ruled: project managers. Unlike his biological counterpart, the project manager is more agile and adaptable than the beast he's displacing, more likely to live by his wits than throwing his weight around.

Effective project managers will remain an indispensable commodity for successful organizations in the coming years. More and more companies are coming to this conclusion and adopting project management as a way of life. Indeed, companies in such diverse industries as construction, heavy manufacturing, insurance, health care, finance, public utilities, and software are becoming project savvy and expecting their employees to do the same.

Although there are a number of general definitions of the term project, we must recognize at the outset that projects are distinct from other organizational processes. As a rule, a process refers to ongoing, day-to-day activities in which an organization engages while producing goods or services. Processes use existing systems, properties, and capabilities in a continuous, fairly repetitive manner. Projects, on the other hand, take place outside the normal, process-oriented world of the firm. Certainly, in some organizations, such as construction, day-to-day processes center on the creation and development of projects. Nevertheless, for the majority of organizations project management activities remain unique and separate from the manner in which more routine, process-driven work is performed. Project work is continuously evolving, establishes its own work rules, and is the antithesis of repetition in the workplace. As a result, it represents an exciting alternative to "business as usual" for many companies. The challenges are great, but so are the rewards of success.

First, we need a clear understanding of the properties that make projects and project management so unique. Consider the following definitions of projects:

A project is a unique venture with a beginning and end, conducted by people to meet established goals within parameters of cost, schedule, and quality.

A project can be considered to be any series of activities and tasks that:
- Have a specific objective to be completed within certain specifications
- Have defined start and end dates

- Have funding limits, if applicable
- Consume human and nonhuman resources, such as money, people, equipment
- Are multifunctional (i.e., cut across several functional lines)

The definition of project management can be found in the Project Management Body of Knowledge (PMBOK ®) Guide of the Project Management Institute (PMI). The PMI is the world's largest professional project management association, with almost 700,000 members worldwide as of 2023. In the PMBOK ® Guide, project management is defined as the "application of knowledge, skills, tools, and techniques to project activities to meet the project requirements". Examples of typical projects include an annual senior management conference, adding multiple languages to user manuals, the development of software, and improving manufacturing cycle-time, to name a few.

While the previous edition of PMBOK was project manager and process focused, the most recent PMBOK is project team and outcome focused. The five process groups of the sixth edition have been replaced by twelve Project Management Principles. These principles "are built around a set of statements that guide the actions and behaviors of project management practitioners regardless of development approach." The twelve principles are:

"Stewardship

Team

Stakeholders

Value

Systems Thinking

Leadership

Tailoring

Quality

Complexity

Risk

Adaptability and Resilience

Change"

The seventh edition of PMBOK ® Guide has also replaced the ten Knowledge Areas with eight Performance Domains. PMI defines a Performance Domain as "groups of related activities that are critical for the effective delivery of project outcomes." The eight domains are:

Stakeholders

Team

Development Approach and Life Cycle

Planning

Project Work

Delivery

Measurement

Uncertainty

These changes reflect the global shift in project management itself. As stated in the "Pulse of the Profession" report, agility and flexibility are required to be successful in today's organizational and project environment.

It would be better if the PMBOK ® Guide specified that a project manager should facilitate planning. One mistake made by inexperienced project managers is to plan the projects for their teams. Not only do they get no buy-in to their plans, but their plans are usually full of holes. Managers can't think of everything, their estimates of task durations are wrong, and everything falls apart after the projects are started. The first rule of project management is that the people who must do the work should help plan it.

The role of the project manager is that of an enabler. Her job is to help the team get the work completed, to "run interference" for the team, to get scarce resources that team members need, and to buffer them from outside forces that would disrupt the work. She is not a project czar. She should be—above all else—a leader, in the truest sense of the word.

The best definition of leadership that I have found is the one by Vance Packard, in his book The Pyramid Climbers (Crest Books, 1962). He says, "Leadership is the art of getting others to want to do something that you believe should be done." The operative word here is "want." Dictators get others to do things that they want done. So do guards who supervise prison work teams. But a leader gets people to want to do the work, and that is a significant difference.

The planning, scheduling, and control of work represent the management or administrative parts of the job. But, without leadership, projects tend to just satisfy bare minimum requirements. With leadership, they can exceed those bare minimums.

Notes

(1) competitive niche 竞争性细分市场:指市场上的特定细分部分,供应商通过满足特定需求或目标群体来与竞争对手区分开来,并获得竞争优势。

(2) stewardship:管家式管理

(3) tailoring:定制化

(4) run interference:做好后勤、作掩护阻挡

参考译文

第9课　项目和项目管理

项目是我们改变世界的主要手段之一。无论目标是分裂原子、挖掘穿越英吉利海峡的隧道、推出 Windows 10 还是筹划奥运会,实现这些挑战的手段始终是相同的:项目管理。

项目管理已经成为最受欢迎的工具之一,通过项目管理,公众或私人组织可以改善内部运作,快速响应外部机遇,取得技术突破,简化新产品开发,从而更有效应对商业环境中出现的挑战。畅销书作家和管理咨询师汤姆·彼得斯指出项目管理及其在企业中的地位:"与重复性任务相比,项目现在是大多数企业中创造增值的基础。"项目管理已成为全球组织成功经营的关键组成部分之一。

现代企业的一个关键特点就是具有外部事件带来的机遇和威胁。企业从来没有像现在这样,既要面对国际竞争,同时还需要迅速抓住商业机会。公司必须不断修改和推出新产品,对客户的需求迅速做出反应,同时还要保持竞争成本和经营水平。完成所这些任务是不是不可能?以前确实不可能。传统的看法认为企业可以通过低成本策略、产品创新或对注重客户服务来参与竞争。简而言之,也就是企业不得不选择其有竞争力的细分市场,而将其他市场份额让给他人。然而,在过去20年中,一切都变了。通用电气、苹果、爱立信、波音和甲骨文等公司变得越来越擅长实现所有这些目标,而不只是满足其中之一。这些公司似乎在竞争模式的每个方面都取得了成功:它们快速进入市场并高效运作,注重成本,关注客户。它们是如何完成这些不可能的事情的呢?

显然这个复杂问题不止一个答案。但毫无疑问的是,这些公司至少有一个共同特征:它们开发项目管理并致力于将其作为竞争手段。《财富》杂志曾经有过这样的描述:

老一代的中层管理者如今已经过时,"新一代项目经理人"正在演变以填补他们曾经统治的空缺。与老一代人不同,项目经理比他所取代的中层管理者更灵活,适应性更强,更可能凭借自己的智慧生存,而不是占着位置仗势欺人。

在未来几年中,高效的项目经理将继续成为成功组织不可或缺的宝贵资源。越来越多的公司得出这一结论并将项目管理作为一种企业生存的方式。实际上,建筑、重工制造、保险、医疗、金融、公用事业和软件等各行各业的公司已经对项目越来越了解,并期望他们的员工也这样做。

虽然对项目这个术语有许多一般性的定义,但我们必须首先认识到项目(project)与其他组织流程的不同。流程(process)通常指的是持续的日常活动,组织利用这些活动来生产产品或提供服务,流程以连续重复的方式对现有的系统、资产和能力进行利用。而项目产生于企业常规的以流程为导向的领域之外。尽管某些组织中,比如建筑行业,日常流程着重于项目设计和开发。然而对于大多数组织来说,项目管理活动都是独特的,区别于其他以流程为导向的日常活动。项目工作持续发展,有着自己的工作规则,但又不是工作场所重复性的工作。因此,项目为"一切照常"的许多企业提供了令人兴奋的替代物。挑战很大,但成功的回报也是巨大的。

首先,要清楚地了解使项目和项目管理如此独特的属性。看看以下关于项目的定义:

项目是具有开始和结束的一次性独特的努力,由相关人员执行以达到符合一定成本、预算和质量要求的目标。

项目可以被视为任何一系列的活动和任务,它们:
- 有特定的目标,需要根据一定的规范完成
- 有明确的开始和结束日期
- 有资金限制(如果适用)
- 消耗人力和非人力资源(如资金、人员、设备等)

● 是多功能的(如涉及多个职能部门)

项目管理的定义可在(美国)项目管理协会(Project Management Institute,简称 PMI)的《项目管理知识体系指南》(Project Management Body of Knowledge,简称 PMBOK ® Guide)中找到。(美国)项目管理协会是世界上最大的专业项目管理协会,2023 年全球会员人数接近 70 万人。在《项目管理知识体系指南》中,项目管理被定义为"应用知识、技能、工具和技术于项目活动以满足项目需求"。典型项目的示例包括年度高级管理会议、为用户手册添加多种语言、软件开发以及改进制造周期等。

相对之前的《项目管理知识体系指南》版本更加侧重于项目经理和过程,最新的版本更注重项目团队和结果。第六版的五个过程组被十二个项目管理原则所取代。这些原则"基于一套指导项目管理从业人员的行为的陈述,无论开发方法为何"。这十二个原则包括:

管家式管理
团队
干系人
价值
系统思考
领导力
定制化
质量
复杂性
风险
适应性和韧性
变革

第七版 PMBOK 还用八个项目绩效领域替代了十个知识领域。PMI 将绩效领域定义为"与项目结果的有效交付相关的几组相关活动"。这八个领域包括:

干系人
团队
开发方法和生命周期
规划
项目工作
交付
测量
不确定性

这些变化反映了项目管理本身的全球转变。正如《专业的脉动》报告中所述,敏捷性和灵活性是在当今组织和项目环境中取得成功所必需的。

《项目管理知识体系指南》若明确指出"项目经理应促成规划"就更好了。缺乏经验的项目经理常犯的一个错误是,替整个团队制定规划。这样制定出的规划不仅不能获得认同,还常常漏洞百出。项目经理不可能面面俱到,比如他们对任务所需时间的估计是错误的,一旦项目启动就会纰漏不断。项目管理的第一条准则就是：真正做事的人务必参与规划。

项目经理的角色应该是"赋能者"，也就是帮助团队完成工作，为团队"做好后勤"，为团队成员获取必需的稀缺资源，为他们屏蔽外力干扰。项目经理首先应是一个名副其实的"领导者"，不应是一个"皇帝"。

目力所及，我认为万斯·帕卡德给"领导"所下的定义堪称最佳。他在《金字塔攀登者》一书中写道："领导是一门艺术，是让他人心甘情愿地去做你认为应该做的事。"这里的关键词是"心甘情愿"，独裁者也能指使他人做事，狱警能监督犯人干活，可是领导者是让人心甘情愿做事，这中间差别很大。

项目工作的规划、排时程以及控制属于管理或行政的部分。但是，如果没有领导力，项目只能达成最低要求。有了领导力之后，就能达成更多要求。

英翻中练习

(1) In the official project management world, too many people focus solely on the project management process when leading a project. And unfortunately, the unofficial project manager follows suit. They both have a blind spot when it comes to leading the people involved in making the project succeed. Their prevailing mindset is this formula:

Process＝Success

A great process is one key to great project success, but the process is only half of the equation. The other half of the formula—equally important as the first half—is leading people:

People＋Process＝Success

The unofficial project manager who adopts the mindset of PEOPLE + PROCESS knows the secret to project success and can replicate it over and over again.

(2) Once upon a time, people didn't think much about the human side of project management. But research has caught up with today's reality. "In the 1980s, we believed that the failure of a project was largely a quantitative failure due to ineffective planning, scheduling, estimating, cost control, and 'moving targets,'" says Harold Kerzner, a top researcher in the field of project management. "During the 1990s, we changed our view of failure from being quantitatively oriented to qualitatively oriented." The thinking now is that failure can be "largely attributed to poor morale, poor motivation, poor human relations, poor productivity, and no employee commitment."

Managing the process with excellence is important, but being a good leader is essential. Enforcing project management techniques can never substitute for motivating and empowering people to implement them themselves. You want people to work with you, not against you. You've somehow got to inspire your team to fully commit to the project and motivate them to follow the process if you're going to achieve long-term success.

Why? Because efficiency and control strategies rarely work—especially in the long run. If you're an official project manager or boss, you may be able to coax people into being productive, but rarely can you force them to bring their most creative energy and efforts to a project. And as an unofficial project manager, you often lack the formal authority to tell anyone what to do. For instance, do the people on your project team report to you? In many cases, they don't. Can you order people to perform? Probably not. To have real project success, you need what we call "informal authority."

Informal authority inspires people to want to play on your team and win.

Formal authority comes from a title or a position. Giving people titles doesn't necessarily make them good leaders. A title may allow someone to enforce rules or penalize team members when rules aren't followed, but titles alone rarely guarantee willing followers who cheerfully volunteer their best talent and effort.

In contrast, informal authority comes from the character and capabilities of a leader. For example, Mahatma Gandhi never held a formal position, yet he led India to independence and inspired movements for civil rights and freedom across the world. Nelson Mandela and Martin Luther King Jr. didn't have official leadership titles, yet from their prison cells they inspired, empowered, and effectively led people.

Each of these leaders earned the right to influence through informal authority because they inspired trust due to their strong character and integrity. Informal authority can be far more powerful than formal authority.

(3) Think of a person you have worked with in the past—a coworker, friend, teacher, or informal leader—who inspired you to give your best. Would you say he or she had informal authority? What was it about the person that made you want to contribute? In what ways did she or he motivate you? By a good example? By listening well? By showing you and others respect? How did that person hold you accountable for results that motivated you to perform?

You may not be leading a social movement like Gandhi or Mandela, but as a project manager you will need to inspire people to want to play on your team, contribute fully, and do their best work. Whether or not you hold a formal position of authority, you can become a true leader if you behave like one. Like the leaders just mentioned, you will need strong character and integrity, and the people you lead must feel respected and heard. As each member of your team will have differing needs, personalities, work styles, and talents, they will need you to be consistent in your approach to their needs and the project process.

第二节 新时代的中国能源发展

本节所选短文均涉及中国近年来在能源电力等领域的发展状况及相关科普知识。共有7篇课文,每篇课文后面附有注释及参考译文,供大家阅读理解及翻译练习时使用。

Text 1 Developing High-Quality Energy in the New Era

China's energy strategy in the new era endeavors to adapt to domestic and international changes and meet new requirements. China will continue to develop high-quality energy to better serve economic and social progress, support the Beautiful China and Healthy China initiatives, and build a clean and beautiful world.

1. The New Energy Security Strategy

In its energy plans for the new era, China has adopted a new strategy featuring Four Reforms and One Cooperation.

– **One reform to improve the energy consumption structure by containing unnecessary consumption.** China is determined to carry out the principle of prioritizing energy conservation, and has tightened the control of total energy consumption and energy use intensity, and enforced energy conservation in all areas of social and economic development. It resolves to adjust its industrial structure. It emphasizes energy conservation in the process of urbanization, and works to develop a green and low-carbon transport system. China encourages hard work and thrift and calls people to conserve energy and work and live with green energy, and move faster towards an energy-saving society.

– **One reform to build a more diversified energy supply structure.** In the direction of green development, China has been vigorously promoting the clean and efficient utilization of fossil energy, prioritizing the development of renewable energy, developing nuclear power in a safe and orderly manner, and raising the proportion of non-fossil energy in the energy supply structure. China has intensified efforts for the exploration and exploitation of oil and gas resources, to increase reserve and production volumes. China has been building the production, supply, storage and sales systems for coal, electricity, oil and gas, while improving energy transportation networks, storage facilities, the emergency response system for energy storage, transportation and peak load management, and enhancing its supply capacity for safer and higher-quality energy.

– **One reform to improve energy technologies to upgrade the industry.** China is implementing the innovation-driven development strategy, building a system that nurtures innovation in green energy technologies, and upgrading energy technologies and equipment in an all-round way. China has strengthened basic research on energy, innovation in generic and disruptive technologies, and original and integrated innovation. China has started to integrate digital, big-data and AI technologies with technologies for clean and efficient energy exploration and exploitation, with a focus on smart energy technologies, to turn these technologies and related industries into new growth drivers for industrial upgrading.

– **One reform to optimize the energy system for faster growth of the energy sector.** China is determined to promote energy market reform, to marketize energy commodities and form a unified and open market with orderly competition. China is furthering energy pricing reform, to create a mechanism in which the market determines the price. China has been working to modernize its law-based energy governance system, developing new models of efficient energy management, and pushing forward reforms to streamline government administration, delegate powers, improve regulation, and upgrade service. It has strengthened planning and policy guidance for the energy sector, and improved the regulatory system of the energy industry.

- **Comprehensive cooperation with other countries to realize energy security in an open environment.** Under the principle of equality and mutual benefit, China is opening its door wider to the world. China promotes green and sustainable energy under the Belt and Road Initiative (BRI), and endeavors to improve energy infrastructure connectivity. China has been an active participant in global energy governance, increasing energy cooperation and exchanges with other countries, and facilitating international trade and investment in the energy sector. China has joined the international community in building a new model of energy cooperation, maintaining energy market stability, and safeguarding common energy security.

2. Guiding Philosophies for Energy Policies in the New Era

- **Putting people first.** China upholds the principle of energy development for the people, by the people and answerable to the people. Its primary goal is to ensure energy supply for people's life and to ensure that the poverty-stricken population have access to electricity. To this end, China has been improving energy infrastructure related to people's life and public services, and has integrated energy development with poverty eradication. China has launched programs on poverty reduction through energy support, which exemplify the fundamental role of energy supply, infrastructure and services in the battle against poverty.

- **Promoting a clean and low-carbon energy.** China embraces the vision of harmonious coexistence between humanity and nature, directing its efforts towards clean and low-carbon energy. China promotes green energy production and consumption, and has improved the relevant structures accordingly. China is increasing the proportion of clean energy and non-fossil energy at the consumption stage, reducing carbon dioxide emissions and pollutant discharge by large margins, and working hard to accelerate its transformation towards green and low-carbon development for the Beautiful China initiative.

- **Ensuring the core status of innovation.** China is focusing on transforming the energy sector through technical advancement. China is actively promoting independent innovation in energy technologies, and increasing sci-tech input in the national energy development. With enterprises playing a key role in innovation, China has been promoting close collaboration along the energy industrial chain between enterprises, universities and research institutes, to reduce technology imports and boost independent innovation so as to develop a new model where innovations of both upper and lower streams well interact and coordinate with each other.

— **Pursuing development through reform.** China will fully leverage the decisive role of the market in allocating energy resources, and ensure the government better play its part in this regard. It is endeavoring to advance market-oriented reform in the competitive areas of the energy sector, further display the role of the market mechanism, and build a high-quality energy market system. China has highlighted the guiding role of its energy strategy and planning, formed a law-based governance system and a regulatory system in the energy industry, and improved the financial and fiscal systems that support green and low-carbon energy transformation. All these measures aim to unlock potential and provide support for quality growth of the energy sector.

— **Building a global community of shared future.** Confronted by the severe impact of climate change, China advocates a global community of shared future, greater international cooperation on energy governance, and a new round of energy reform directed towards clean and low-carbon development. China has joined other countries in seeking sustainable energy and building a clean and beautiful world.

Notes

(1) energy use intensity 能耗强度
(2) diversified energy supply structure 多元供应体系
(3) generic and disruptive technologies 共性技术和颠覆性技术
(4) streamline government administration 精简政府行政机构
(5) energy infrastructure connectivity 能源基础设施互联互通

参考译文

第1课 走新时代能源高质量发展之路

新时代的中国能源发展，积极适应国内国际形势的新发展新要求，坚定不移走高质量发展新道路，更好服务经济社会发展，更好服务美丽中国、健康中国建设，更好推动建设清洁美丽世界。

（一）能源安全新战略

新时代的中国能源发展，贯彻"四个革命、一个合作"能源安全新战略。

——推动能源消费革命，抑制不合理能源消费。坚持节能优先方针，完善能源消费总量管理，强化能耗强度控制，把节能贯穿于经济社会发展全过程和各领域。坚定调整产业结构，高度重视城镇化节能，推动形成绿色低碳交通运输体系。在全社会倡导勤俭节约的消费观，培育节约能源和使用绿色能源的生产生活方式，加快形成能源节约型社会。

——推动能源供给革命，建立多元供应体系。坚持绿色发展导向，大力推进化石能源清洁高效利用,优先发展可再生能源,安全有序发展核电,加快提升非化石能源在能源供应中的比重。大力提升油气勘探开发力度,推动油气增储上产。推进煤电油气产供储销体系建设,完善能源输送网络和储存设施,健全能源储运和调峰应急体系,不断提升能源供应的质量和安全保障能力。

——推动能源技术革命,带动产业升级。深入实施创新驱动发展战略,构建绿色能源技术创新体系,全面提升能源科技和装备水平。加强能源领域基础研究以及共性技术、颠覆性技术创新,强化原始创新和集成创新。着力推动数字化、大数据、人工智能技术与能源清洁高效开发利用技术的融合创新,大力发展智慧能源技术,把能源技术及其关联产业培育成带动产业升级的新增长点。

——推动能源体制革命,打通能源发展快车道。坚定不移推进能源领域市场化改革,还原能源商品属性,形成统一开放、竞争有序的能源市场。推进能源价格改革,形成主要由市场决定能源价格的机制。健全能源法治体系,创新能源科学管理模式,推进精简行政机构和"放管服"改革,加强规划和政策引导,健全行业监管体系。

——全方位加强国际合作,实现开放条件下能源安全。坚持互利共赢、平等互惠原则,全面扩大开放,积极融入世界。推动共建"一带一路"能源绿色可持续发展,促进能源基础设施互联互通。积极参与全球能源治理,加强能源领域国际交流合作,畅通能源国际贸易、促进能源投资便利化,共同构建能源国际合作新格局,维护全球能源市场稳定和共同安全。

(二)新时代能源政策理念

——坚持以人民为中心。牢固树立能源发展为了人民、依靠人民、服务人民的理念,把保障和改善民生用能、贫困人口用能作为能源发展的优先目标,加强能源民生基础设施和公共服务能力建设,提高能源普遍服务水平。把推动能源发展和脱贫攻坚有机结合,实施能源扶贫工程,发挥能源基础设施和能源供应服务在扶贫中的基础性作用。

——坚持清洁低碳导向。树立人与自然和谐共生理念,把清洁低碳作为能源发展的主导方向,推动能源绿色生产和消费,优化能源生产布局和消费结构,加快提高清洁能源和非化石能源消费比重,大幅降低二氧化碳排放强度和污染物排放水平,加快能源绿色低碳转型,建设美丽中国。

——坚持创新核心地位。把提升能源科技水平作为能源转型发展的突破口,加快能源科技自主创新步伐,加强国家能源战略科技力量,发挥企业技术创新主体作用,推进产学研深度融合,推动能源技术从引进跟随向自主创新转变,形成能源科技创新上下游联动的一体化创新和全产业链协同技术发展模式。

——坚持以改革促发展。充分发挥市场在资源配置中的决定性作用,更好发挥政府作用,深入推进能源行业竞争性环节市场化改革,发挥市场机制作用,建设高标准能源市场体系。加强能源发展战略和规划的导向作用,健全能源法治体系和全行业监管体系,进一步完善支持能源绿色低碳转型的财税金融体制,释放能源发展活力,为能源高质量发展提供支撑。

——坚持推动构建人类命运共同体。面对日趋严峻的全球气候变化形势,树立人类命运共同体意识,深化全球能源治理合作,加快推动以清洁低碳为导向的新一轮能源变革,共同促进全球能源可持续发展,共建清洁美丽世界。

英翻中练习

(1) Energy is the foundation and driving force for the progress of human civilization. It matters to the economy, to people's lives, to national security, and to the survival and development of humanity. It is of vital importance in advancing social and economic development and public welfare.

Since the founding of the People's Republic of China (PRC) in 1949, under the leadership of the Communist Party of China (CPC), a relatively complete energy industry system has been established. This has largely been achieved through self-reliance and hard work. Since the launch of the reform and opening-up policy in 1978, to adapt to the rapid development of the economy and society, China has promoted the development of energy in a comprehensive, coordinated and sustainable manner. Today, China has become the world's largest energy producer and consumer. Its transition to efficient energy utilization has been the fastest in the world.

Since the 18th CPC National Congress in 2012, China has entered a new era, as has its energy development. In 2014, President Xi Jinping put forward a new energy security strategy featuring Four Reforms and One Cooperation, pointing out the direction for the quality growth of the energy industry with Chinese characteristics in the new era. China upholds the vision of innovative, coordinated, green, open and shared development with focus on high quality and restructuring of the supply side. It has been working on all fronts to reform the ways energy is consumed, to build a clean and diversified energy supply system, to implement an innovation-driven energy strategy, to further the reform of the energy system, and to enhance international energy cooperation. China has entered a stage of high-quality energy development.

(2) A thriving civilization calls for a good eco-environment. Facing increasingly severe global problems such as climate change, environmental risks and challenges, and energy and resource constraints, China embraces the vision of a global community of shared future and accelerates its transformation towards green and low-carbon development in economy and society. In addition to promoting clean and low-carbon energy use domestically, China has been an active participant in global energy governance, exploring a path of worldwide sustainable energy alongside other countries. At the general debates of the 75th United Nations General Assembly in September 2020, President Xi pledged that China will scale up its Intended Nationally Determined Contributions by adopting more vigorous policies and measures, striving to have carbon dioxide emissions peak before 2030 and to achieve carbon neutrality before 2060. In the new era, China's energy strategy will provide forceful support for sound and sustained economic and social development, and make a significant contribution to ensuring world energy security, addressing global climate change, and boosting global economic growth.

Text 2 Historic Achievements in Energy Development

China is committed to driving an energy revolution. As a result, major changes have taken place in the production and use of energy and historic achievements have been realized in energy development. Energy production and consumption are being optimized, energy efficiency has increased significantly, and energy use has become more convenient for both work and life. China's capacity to ensure energy security has been strengthened. All this provides important support to quality economic development, victory in the battle against poverty, and building a moderately prosperous society in all respects.

1. Growing Capacity to Ensure Energy Supply

A diversified energy production infrastructure consisting of coal, oil, natural gas, electricity, nuclear energy, new energy and renewable energy is in place. Preliminary calculations show that China's primary energy production in 2019 reached 3.97 billion tons of standard coal, making it the world's largest energy producer.

Coal remains the basic energy source. Since 2012, the annual production of raw coal has ranged between 3.41 and 3.97 billion tons. Crude oil production remains stable. Since 2012, the annual production of crude oil has ranged between 190 and 210 million tons. The production of natural gas has increased notably, from 110.6 billion cubic meters in 2012 to 176.2 billion cubic meters in 2019. China's electricity supply capacity has risen to a cumulative installed capacity of 2.01 billion kW in 2019, up 75 percent since 2012, and an electricity output of 7.5 trillion kWh, up 50 percent. Renewable energy resources have expanded rapidly, with cumulative installed capacities of hydropower, wind power, and solar photovoltaic (PV) power each ranking top in the world. As of the end of 2019, the total installed capacity of nuclear power plants under construction and in operation reached 65.93 million kW, the second largest in the world. The installed capacity of nuclear power plants under construction ranked first.

Energy transport capacity has risen remarkably. China has built natural gas trunk lines measuring over 87,000 km, oil trunk lines totaling 55,000 km, and 302,000 km of electricity transmission lines of 330 kv or more.

The energy reserve system has been steadily improved. China has built nine national oil reserve bases; it has achieved preliminary results in building a natural gas production, supply, reserve and sale system; the coordinated guarantee system for coal production and transport is sound; the country has become a global leader in operating a secure and stable power grid; and its capacity in comprehensive energy emergency response has been strengthened significantly.

2. Remarkable Achievements in Optimizing Energy Conservation and Consumption

Significant improvement has been made in energy efficiency. Since 2012, energy

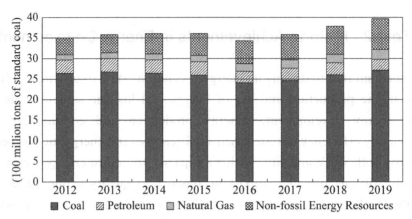

Figure 1　China's Energy Production(2012～2019)
Source: National Bureau of Statistics

consumption per unit of GDP has been reduced by 24.4 percent, equivalent to 1.27 billion tons of standard coal. From 2012 to 2019, China saw an average annual growth of 7 percent in the economy, while annual energy consumption rose by only 2.8 percent.

The shift towards clean and low-carbon energy consumption is accelerating. Preliminary calculations show that in 2019, coal consumption accounted for 57.7 percent of total energy consumption, a decrease of 10.8 percentage points from 2012; the consumption of clean energy (natural gas, hydropower, nuclear power, wind power) accounted for 23.4 percent of total energy consumption, an increase of 8.9 percentage points over 2012. Non-fossil energy accounted for 15.3 percent of total energy consumption, up 5.6 percentage points against 2012. With this China has reached the target of raising the share of non-fossil energy to 15 percent in total energy consumption by 2020. The number of new energy vehicles is rising rapidly. In 2019 the total number of new energy vehicles reached 3.8 million, with 1.2 million new energy vehicles going on road that year. Both of these figures represent more than half of the global totals. As of the end of 2019, there were 1.2 million electric-vehicle charging stations nationwide, constituting the largest charging network in the world, and effectively improving energy efficiency and optimizing energy consumption in the transport sector.

3. Rapid Improvements in Energy Technology

China continues to pursue technological innovation in the energy sector. Its energy technologies are continuously improving, and technological progress has become a basic driver for the transformation of the energy industry. There are complete industrial chains for the manufacturing of clean energy equipment for hydropower, nuclear power, wind power, and solar power. China has successfully developed and manufactured the world's largest single-unit hydropower generators, with a capacity of 1 million kW; it is able to manufacture a full range of wind turbines with a maximum single-unit capacity of 10 mW; and it continues to establish new world bests in the conversion efficiency of solar

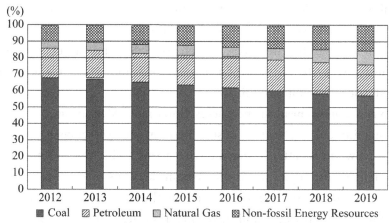

Figure 2 China's Energy Consumption Structure(2012～2019)
Source: National Bureau of Statistics

PV cells. China has built a number of nuclear power plants using advanced third-generation technologies, and made significant breakthroughs in a number of nuclear energy technologies such as new-generation nuclear power generation and small modular reactors. Its technological capabilities in oil and gas exploration and development keep improving. It leads the world in technologies such as the high-efficiency development of low-permeability crude oil and heavy oil, and a new generation of compound chemical flooding. The technology and equipment for shale oil and gas exploration and development have greatly improved, and successful natural gas hydrate production tests have been completed. China is developing green, efficient and intelligent coal mining technology. It has achieved mechanization in 98 percent of its large coal mines, and mastered the technology for producing oil and gas from coal. It has built a safe, reliable, and world-leading power grid which is the largest across the globe, with reliability of supply at the forefront of the world. A large number of new energy technologies, new businesses, and new models such as "Internet +" smart energy, energy storage, block chain, and integrated energy services are booming.

4. Significant Progress in Eco-Environmental Friendliness of the Energy Sector

China sees green energy as an important measure to enhance eco-environmental progress, and resolutely fights pollution, especially air pollution. Its capabilities in clean coal mining and utilization have greatly improved, and significant results have been achieved in regulating coal mining subsidence areas and building green mines. It has amended the *Law on Air Pollution Prevention and Control* to strengthen the prevention and control of pollution from coal and other energy sources, and ensure that more environmentally friendly energy sources are used to replace coal in equal or reduced amount in newly-built, renovated, or expanded coal-consuming projects in key areas for air pollution control. The green development of the energy sector has significantly improved air quality, and the emissions of sulfur dioxide, nitrogen oxides and soot have dropped notably. Green development of the energy sector has played an important role in reducing carbon emissions. By 2019, carbon emission intensity in China had decreased by 48.1 percent compared with 2005, which exceeded the target of reducing carbon emission intensity by 40 to 45 percent between 2005 and 2020, reversing the trend of rapid carbon dioxide emission growth.

5. Continuous Improvement in the Energy Governance Mechanism

China is making every effort to ensure that the market plays a greater role in the energy sector. Now, in a better business environment and a more viable market, market entities and individuals enjoy more convenient services and find it easier to start businesses. Market access for foreign capital in the energy sector has been extended, private investment is growing, and investment entities have become more diverse. Policies on power generation and consumption plans have been relaxed in an orderly manner, trading institutions can operate independently and in accordance with regulations, and the power market has further developed. China has accelerated reforms such as the deregulation of the oil and gas exploration market and the circulation of mining rights, reform of the pipeline network operation mechanism, and the dynamic management of crude oil imports. It has improved the construction of oil and gas trading centers. China encourages the market to play a decisive role in determining energy prices. It has further relaxed control on the prices in competitive areas, and has preliminarily established a reasonable pricing mechanism for power transmission and distribution and oil and gas pipeline networks. It coordinates energy reform with law-based governance, and the legal framework regarding the energy sector has been improved. An energy governance mechanism covering strategies, plans, policies, standards, supervision, and services is in place.

6. Solid Benefits for People's Lives

Ensuring public wellbeing and improving people's lives is China's fundamental goal in energy development. China is ensuring that urban and rural residents have access to basic energy supply and services, as a fundamental element in building a moderately prosperous society in all respects and supporting rural revitalization. From 2016 to 2019, the total investment in transforming and upgrading rural power grids reached RMB830 billion, and the average power outage time in rural areas was reduced to about 15 hours per year. The quality of power services for rural residents has improved significantly. From 2013 to 2015 China implemented an action plan to ensure access to electricity for every citizen, and completed this historic task by the end of 2015. It has implemented poverty alleviation projects based on solar PV power generation, and other energy-related poverty alleviation projects. China prioritizes poverty-stricken areas in planning energy development projects and has introduced energy projects for the benefit of the rural residents. This has promoted economic development in poverty-stricken areas and raised the incomes of the poor. It has improved the infrastructure for natural gas utilization, supplied natural gas to more areas, and improved its ability to ensure gas supply for people's daily life. Significant progress has been made in clean heating in northern China, with improvements in the energy use and living environment of urban and rural residents. As of the end of 2019, clean heating in northern China covered a floor space of 11.6 billion sq m, an increase of 5.1 billion sq m over 2016.

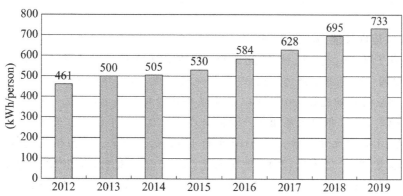

Figure 3　China's Per Capita Household Electricity Consumption(2012～2019)
Source: China Electricity Council

Notes

(1) primary energy production 一次能源生产：指未经转换直接从自然资源中获得的能源，如煤炭、石油、天然气等。

(2) trunk line 干线：高电压输电线路的主要干线，用于电力长距离传输。

(3) energy reserve system 能源储备体系：指国家建立的能源储备和供应保障体系。

(4) charging station 充电站：提供电动车辆充电设施的场所。

(5) compound chemical flooding 化合物化学驱油：一种增强油田采收率的技术，通过注入化学物质来改变油藏物理和化学特性。

(6) carbon emission intensity 碳排放强度：单位产出或单位活动所排放的二氧化碳数量。

(7) market entity 市场主体：指在市场经济中参与经济活动并享有权益的个人、企业或组织。

(8) deregulation 取消管制：降低或取消对特定行业或经济活动的政府管制与限制。

(9) circulation of mining rights 矿业权流转：指矿业权所有人之间对矿业权进行转让、交易的过程。

(10) rural revitalization 乡村振兴：促进农村经济发展和改善农民生活质量的政策或行动。

(11) clean heating 清洁供暖：利用清洁能源或高效能源实现低污染、低排放的供暖方式。

参考译文

第2课 能源发展取得历史性成就

中国坚定不移推进能源革命,能源生产和利用方式发生重大变革,能源发展取得历史性成就。能源生产和消费结构不断优化,能源利用效率显著提高,生产生活用能条件明显改善,能源安全保障能力持续增强,为服务经济高质量发展、打赢脱贫攻坚战和全面建成小康社会提供了重要支撑。

1. 能源供应保障能力不断增强

基本形成了煤、油、气、电、核、新能源和可再生能源多轮驱动的能源生产体系。初步核算,2019年中国一次能源生产总量达39.7亿吨标准煤,为世界能源生产第一大国。煤炭仍是保障能源供应的基础能源,2012年以来原煤年产量保持在34.1亿—39.7亿吨。努力保持原油生产稳定,2012年以来原油年产量保持在1.9亿—2.1亿吨。天然气产量明显提升,从2012年的1106亿立方米增长到2019年的1762亿立方米。电力供应能力持续增强,累计发电装机容量20.1亿千瓦,2019年发电量7.5万亿千瓦时,较2012年分别增长75%、50%。可再生能源开发利用规模快速扩大,水电、风电、光伏发电累计装机容量均居世界首位。截至2019年底,在运在建核电装机容量6593万千瓦,居世界第二,在建核电装机容量世界第一。

能源输送能力显著提高。建成天然气主干管道超过8.7万公里、石油主干管道5.5万公里、330千伏及以上输电线路长度30.2万公里。

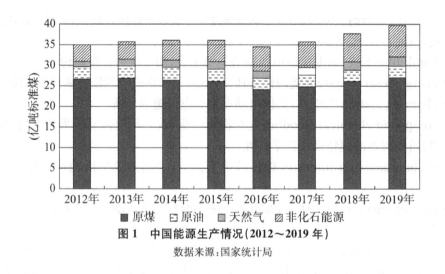

图1 中国能源生产情况(2012~2019年)

数据来源:国家统计局

能源储备体系不断健全。建成9个国家石油储备基地,天然气产供储销体系建设取得初步成效,煤炭生产运输协同保障体系逐步完善,电力安全稳定运行达到世界先进水平,能源综合应急保障能力显著增强。

2. 能源节约和消费结构优化成效显著

能源利用效率显著提高。2012年以来单位国内生产总值能耗累计降低24.4%,相当于减少能源消费12.7亿吨标准煤。2012年至2019年,以能源消费年均2.8%的增长支撑了国民经济年均7%的增长。

能源消费结构向清洁低碳加快转变。初步核算,2019年煤炭消费占能源消费总量比重为57.7%,比2012年降低10.8个百分点;天然气、水电、核电、风电等清洁能源消费量占能源消费总量比重为23.4%,比2012年提高8.9个百分点;非化石能源占能源消费总量比重达15.3%,比2012年提高5.6个百分点,已提前完成到2020年非化石能源消费比重达到15%左右的目标。新能源汽车快速发展,2019年新增量和保有量分别达120万辆和380万辆,均占全球总量一半以上;截至2019年底,全国电动汽车充电基础设施达120万处,建成世界最大规模充电网络,有效促进了交通领域能效提高和能源消费结构优化。

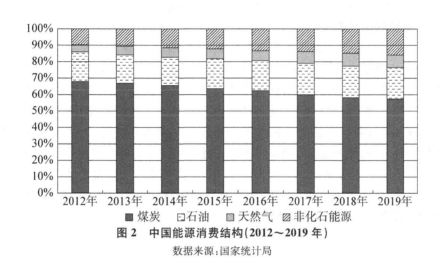

图2　中国能源消费结构(2012～2019年)

数据来源:国家统计局

3. 能源科技水平快速提升

持续推进能源科技创新,能源技术水平不断提高,技术进步成为推动能源发展动力变革的基本力量。建立完备的水电、核电、风电、太阳能发电等清洁能源装备制造产业链,成功研发制造全球最大单机容量100万千瓦水电机组,具备最大单机容量达10兆瓦的全系列风电机组制造能力,不断刷新光伏电池转换效率世界纪录。建成若干应用先进三代技术的核电站,新一代核电、小型堆等多项核能利用技术取得明显突破。油气勘探开发技术能力持续提高,低渗原油及稠油高效开发、新一代复合化学驱油等技术世界领先,页岩油气勘探开发技术和装备水平大幅提升,天然气水合物试采取得成功。发展煤炭绿色高效智能开采技术,大型煤矿采煤机械化程度达98%,掌握煤制油气产业化技术。建成规模最大、安全可靠、全球领先的电网,供电可靠性位居世界前列。"互联网+"智慧能源、储能、区块链、综合能源服务等一大批能源新技术、新模式、新业态正在蓬勃兴起。

4. 能源与生态环境友好性明显改善

中国把推进能源绿色发展作为促进生态文明建设的重要举措,坚决打好污染防治攻坚战、打赢蓝天保卫战。煤炭清洁开采和利用水平大幅提升,采煤沉陷区治理、绿色矿山建设取得显著成效。落实修订后的《大气污染防治法》,加大燃煤和其他能源污染防治力度。推动国家大气污染防治重点区域内新建、改建、扩建用煤项目实施煤炭等量或减量替代。能源绿色发展显著推动空气质量改善,二氧化硫、氮氧化物和烟尘排放量大幅下降。能源绿色发展对碳排放强度下降起到重要作用,2019年碳排放强度比2005年下降48.1%,超过了2020

年碳排放强度比2005年下降40%—45%的目标,扭转了二氧化碳排放快速增长的局面。

5. 能源治理机制持续完善

全面提升能源领域市场化水平,营商环境不断优化,市场活力明显增强,市场主体和人民群众办事创业更加便利。进一步放宽能源领域外资市场准入,民间投资持续壮大,投资主体更加多元。发用电计划有序放开、交易机构独立规范运行、电力市场建设深入推进。加快推进油气勘查开采市场放开与矿业权流转、管网运营机制改革、原油进口动态管理等改革,完善油气交易中心建设。推进能源价格市场化,进一步放开竞争性环节价格,初步建立电力、油气网络环节科学定价制度。协同推进能源改革和法治建设,能源法律体系不断完善。覆盖战略、规划、政策、标准、监管、服务的能源治理机制基本形成。

6. 能源惠民利民成果丰硕

把保障和改善民生作为能源发展的根本出发点,保障城乡居民获得基本能源供应和服务,在全面建成小康社会和乡村振兴中发挥能源供应的基础保障作用。2016年至2019年,农网改造升级总投资达8300亿元,农村平均停电时间降低至15小时左右,农村居民用电条件明显改善。2013年至2015年,实施解决无电人口用电行动计划,2015年底完成全部人口都用上电的历史性任务。实施光伏扶贫工程等能源扶贫工程建设,优先在贫困地区进行能源开发项目布局,实施能源惠民工程,促进了贫困地区经济发展和贫困人口收入增加。完善天然气利用基础设施建设,扩大天然气供应区域,提高民生用气保障能力。北方地区清洁取暖取得明显进展,改善了城乡居民用能条件和居住环境。截至2019年底,北方地区清洁取暖面积达116亿平方米,比2016年增加51亿平方米。

图3 中国人均生活用电量(2012~2019年)

数据来源:中国电力企业联合会

英翻中练习

(1) China's Renewable Energy Exploitation Ranks First in the World

As of the end of 2019, China's total installed capacity of power generation using renewable energy resources reached 790 million kW, accounting for about 30 percent of the global total. The total installed capacity of hydropower reached 356 million kW, wind power 210 million kW, solar PV power 204 million kW, and biomass power 23.69

million kW. All of these ranked first in the world. Since 2010, China has invested a total of about US $ 818 billion in new energy power generation, accounting for 30 percent of the global total investment over the same period.

Heating-supply using renewable energy has been widely adopted. By the end of 2019 the total surface area of solar panels on solar water heaters had reached 500 million sq m. The total floor area of buildings heated using shallow, medium and deep geothermal energy exceeded 1.1 billion sq m.

A complete industrial chain has been formed in the manufacture of wind power and solar PV power generation equipment, with the scale of output and level of technology leading the world. In 2019, the output of polysilicon accounted for 67 percent of the global total. The figure for solar PV cells was 79 percent, and for solar PV modules 71 percent. Solar PV products were exported to more than 200 countries and regions. The production of complete wind power assemblies accounted for 41 percent of the world total, making China a key player in the global industry chain of wind power equipment manufacture.

(2) Achievements in Clean Development of Fossil Energy

Clean coal mining capacity has increased significantly. China is actively promoting clean coal mining technologies such as cut-and-fill mining and water-preserved mining, and strengthening the comprehensive utilization of coal mine resources. In 2019, the coal washing rate reached 73.2 percent, the comprehensive utilization rate of mine water reached 75.8 percent, and the land recovery rate for farming reached 52 percent.

China has built the world's largest clean coal power supply system. It has rolled out ultra-low emission transformation of coal-fired power plants. As of the end of 2019, the total capacity of ultra-low-emission coal power generating units reached 890 million kW, accounting for 86 percent of the total installed capacity of all coal power generating units. Coal-fired power generation units with a total capacity of more than 750 million kW have undergone energy-saving transformation. As a result, the coal consumption of coal-fired power generation has been reduced year by year.

Remarkable results have been achieved in the replacement and transformation of coal-fired furnaces and kilns. Over 200,000 small coal-fired boilers have been phased out, and coal-fired boilers below 35 t/h in key pollution control areas have been basically eliminated. Clean fuel substitution is being implemented for industrial kilns that use coal, petroleum coke, and heavy oil as fuel.

The environmental standards of vehicle fuel have been steadily raised. Through a special campaign to upgrade the quality of refined oil products, the standards of gasoline and diesel for vehicles have been upgraded rapidly, from the National III emission standard in 2012 to the National VI emission standard in 2019, significantly reducing vehicle exhaust emissions.

Text 3 An All-Round Effort to Reform Energy Consumption

China perseveres with its fundamental national policy of conserving resources and protecting the environment. Prioritizing energy saving, it understands that energy conservation means increasing resources, reducing pollution, and benefiting humanity, and exercises energy saving throughout the whole process and in all areas of economic and social development.

1. Implementing a Dual Control System of Total Energy Consumption and Energy Intensity

A dual control system of total energy consumption and energy intensity is in place. China sets the targets of total energy consumption and energy intensity for different provinces, autonomous regions and municipalities directly under the central government and applies oversight and checks over the performance of local governments at all levels. It has introduced the energy-saving index into the performance evaluation system of eco-environmental progress and green development, to guide the transformation of the development philosophy. It breaks down the dual control targets of total energy consumption and energy intensity for key energy consumers, and evaluates their performance accordingly to strengthen energy-saving management.

2. Improving Laws, Regulations and Standards for Energy Conservation

China has revised the *Energy Conservation Law.* It has put in place an energy-saving system in key areas including industry, construction and transport as well as in public institutions. It continues to improve the supporting legal institutions for energy conservation supervision, energy-efficiency labeling, energy-saving checks on fixed assets investment projects, and energy conservation management of key energy consumers. It has strengthened standard-setting as a constraining factor and improved energy-saving standards system. It has carried out 100 projects to upgrade energy efficiency standards, enacted more than 340 national energy-saving standards, including almost 200 mandatory standards, covering most high energy-consuming industries and final energy consumption products. China has strengthened oversight over energy-saving law enforcement, reinforced operational and post-operational supervision, and exercised strict accountability for law enforcement to ensure the effective implementation of energy conservation laws, regulations, and mandatory standards.

3. Improving Energy-Saving and Low-Carbon Incentives

Corporate income tax and value-added tax incentives are awarded to energy-saving businesses. China encourages the imports of energy-saving technologies and equipment, and controls the exports of energy-intensive and heavy-polluting products. China is improving the green financial system, and makes use of energy efficiency credits and green bonds to support energy conservation projects. It is exploring new ground in pricing to advance green development. Differential pricing, time-of-use pricing, and tiered pricing for electricity and natural gas have been adopted. China is improving its policies of environment-friendly electricity pricing to arouse the enthusiasm of market entities and the public in energy conservation. It has conducted trials of paid use of and trading in energy-using right in four provinces and cities including Zhejiang, and carbon emissions trading in seven provinces and cities including Beijing. The government is promoting energy performance contracting (EPC) and developing integrated energy services, and encourages innovations in energy-saving technology and business models. It has strengthened the management of demand-side power use and implemented a market response mechanism to guide the economical, orderly and rational utilization of electricity. A "leader board" of best energy-savers has been put in place to increase the efficiency of final energy consumption products, energy-intensive industries, and public institutions.

4. Improving Energy Efficiency in Key Areas

China is doing all it can to optimize the industrial structure, develop advanced manufacturing, high-tech industry and modern services with low energy consumption, and promote the intelligent and clean transformation of traditional industries. China has sped up the transformation to green, recycling and low-carbon industry, and implemented green manufacturing on all fronts; put in place monitoring, law enforcement and diagnostic mechanisms for energy conservation, and carried out energy efficiency benchmarking; raised the energy-saving standards of new buildings, expanded the energy-saving renovation of existing buildings, and improved the structure of energy consumption in construction. It is developing a highly efficient and comprehensive transport system with lower energy consumption, promoting the use of clean energy in transport, and enhancing energy efficiency of vehicles and other means of transport. It is building energy-saving public institutions, to set an example for the rest of society. A market-oriented system of green technology innovation will be put in place to encourage the R&D, transfer and popularization of green technology. China is promoting national key energy-saving and low-carbon technologies, particularly for the transport sector, and energy-saving industrial equipment. The government encourages extensive public involvement in energy conservation, and is raising public awareness of frugality, promoting simple, modest, green and low-carbon lifestyles, and opposing extravagance and excessive consumption.

5. Promoting Clean Final Energy Consumption

Focusing on the Beijing-Tianjin-Hebei region and its surrounding areas, the Yangtze River Delta, the Pearl River Delta and the Fenwei Plain (the Fenhe Plain, the Weihe Plain and their surroundings in the Yellow River Basin), China is working to reduce and find substitutes for coal consumption, and taking comprehensive measures to control the use of bulk coal. It is promoting clean and efficient coal-fired furnaces, and replacing inefficient and highly-polluting coal with natural gas, electricity and renewable energy. Now, fiscal and price policies are in place to support clean heating in winter in northern China to improve air quality. China is replacing coal and oil with electricity in final energy consumption, and popularizing new energy vehicles, heat pumps, electric furnaces, and other new forms of energy consumption. It has strengthened the development and connectivity of natural gas infrastructure, and made the use of natural gas more efficient in urban areas, as well as in industrial fuel, power generation, and transport. It is promoting natural gas CCHP (combined cooling heating and power), decentralized renewable energy, and multi-energy coordination and energy cascade use in final energy consumption.

Notes

（1）dual control system of total energy consumption and energy intensity 能源消耗总量和能源强度双控制系统

（2）fixed assets investment project 固定资产投资项目

（3）energy efficiency credits 能效信贷：用于支持节能项目的信用贷款。

（4）time-of-use pricing 峰谷分时计价：电力或天然气按使用时段不同而定价。

（5）tiered pricing for electricity 阶梯电价

（6）energy performance contracting (EPC) 能源绩效合同：以节约能源为目标的合同。

（7）demand-side power use 需求侧用电：指电力需求方的用电行为。

（8）energy efficiency benchmarking 能效基准、能效对标：将能源效率作为评估标准进行比较。

（9）Fenwei Plain 汾渭平原：黄河流域汾河平原、渭河平原及其周边台塬阶地的总称。

（10）CCHP (combined cooling heating and power) 热电冷联供的供能系统、热电冷三联供：联合供冷供热和发电的系统。

参考译文

第3课　全面推进能源消费方式变革

坚持节约资源和保护环境的基本国策，坚持节能优先方针，树立节能就是增加资源、减少污染、造福人类的理念，把节能贯穿于经济社会发展全过程和各领域。

1. 实行能耗双控制度

实行能源消费总量和强度双控制度，按省、自治区、直辖市行政区域设定能源消费总量和强度控制目标，对各级地方政府进行监督考核。把节能指标纳入生态文明、绿色发展等绩效评价指标体系，引导转变发展理念。对重点用能单位分解能耗双控目标，开展目标责任评价考核，推动重点用能单位加强节能管理。

2. 健全节能法律法规和标准体系

修订实施《节约能源法》，建立完善工业、建筑、交通等重点领域和公共机构节能制度，健全节能监察、能源效率标识、固定资产投资项目节能审查、重点用能单位节能管理等配套法律制度。强化标准引领约束作用，健全节能标准体系，实施百项能效标准推进工程，发布实施340多项国家节能标准，其中近200项强制性标准，实现主要高耗能行业和终端用能产品全覆盖。加强节能执法监督，强化事中事后监管，严格执法问责，确保节能法律法规和强制性标准有效落实。

3. 完善节能低碳激励政策

实行促进节能的企业所得税、增值税优惠政策。鼓励进口先进节能技术、设备，控制出口耗能高、污染重的产品。健全绿色金融体系，利用能效信贷、绿色债券等支持节能项目。创新完善促进绿色发展的价格机制，实施差别电价、峰谷分时电价、阶梯电价、阶梯气价等，完善环保电价政策，调动市场主体和居民节能的积极性。在浙江等4省市开展用能权有偿使用和交易试点，在北京等7省市开展碳排放权交易试点。大力推行合同能源管理，鼓励节能技术和经营模式创新，发展综合能源服务。加强电力需求侧管理，推行电力需求侧响应的市场化机制，引导节约、有序、合理用电。建立能效"领跑者"制度，推动终端用能产品、高耗能行业、公共机构提升能效水平。

4. 提升重点领域能效水平

积极优化产业结构，大力发展低能耗的先进制造业、高新技术产业、现代服务业，推动传统产业智能化、清洁化改造。推动工业绿色循环低碳转型升级，全面实施绿色制造，建立健全节能监察执法和节能诊断服务机制，开展能效对标达标。提升新建建筑节能标准，深化既有建筑节能改造，优化建筑用能结构。构建节能高效的综合交通运输体系，推进交通运输用能清洁化，提高交通运输工具能效水平。全面建设节约型公共机构，促进公共机构为全社会节能工作作出表率。构建市场导向的绿色技术创新体系，促进绿色技术研发、转化与推广。推广国家重点节能低碳技术、工业节能技术装备、交通运输行业重点节能低碳技术等。推动全民节能，引导树立勤俭节约的消费观，倡导简约适度、绿色低碳的生活方式，反对奢侈浪费和不合理消费。

5. 推动终端用能清洁化

以京津冀及周边地区、长三角、珠三角、汾渭平原等地区为重点，实施煤炭消费减量替代和散煤综合治理，推广清洁高效燃煤锅炉，推行天然气、电力和可再生能源等替代低效和高污染煤炭的使用。制定财政、价格等支持政策，积极推进北方地区冬季清洁取暖，促进大气环境质量改善。推进终端用能领域以电代煤、以电代油，推广新能源汽车、热泵、电窑炉等新型用能方式。加强天然气基础设施建设与互联互通，在城镇燃气、工业燃料、燃气发电、交通运输等领域推进天然气高效利用。大力推进天然气热电冷联供的供能方式，推进分布式可再生能源发展，推行终端用能领域多能协同和能源综合梯级利用。

> 英翻中练习

(1) Growing Momentum for Energy Conservation in Key Areas

Strengthening energy conservation in the industrial sector. China has tightened supervision of energy conservation over major national industrial projects, conducted diagnosis of industrial energy saving, and promoted industrial energy conservation and standardization of green manufacturing and growth. A "leader board" of energy efficiency has been introduced for selecting model enterprises in 12 key industries, including steel and electrolytic aluminum. China has taken action to manage the demand-side industrial power users, issued the "Guidelines for Power Demand-side Management in the Industrial Sector", and selected 153 industrial enterprises and parks as models. It is also cultivating energy service integrators in a bid to integrate modern energy services and industrial manufacturing.

Enhancing energy conservation in the construction sector. Building energy efficiency standards are being rigorously enforced in new construction projects. China has piloted ultra-low and near-zero energy consumption buildings, and undertaken energy-saving renovation of existing residential buildings. It is improving energy efficiency in public buildings, and applying renewable energy in construction. By the end of 2019, 19.8 billion sq m of energy-efficient buildings had been erected, accounting for more than 56 percent of existing buildings in urban areas. In 2019, the floor area of new energy-efficient buildings in urban areas exceeded 2 billion sq m.

Promoting energy conservation in transport. China has improved public transport services and promoted multimodal transport. It has built or renovated more electric railway lines, popularized natural gas vehicles (NGVs), developed energy-saving and new energy vehicles, and improved facilities for battery charge and replacement and hydrogen fuel. Docked ships and civil aircraft are encouraged to use shore power, and CNG (compressed natural gas) filling stations and LNG (liquefied natural gas) fueling stations have been built for NGVs. Obsolete energy-wasting vehicles and ships have been phased out. By the end of 2019, China had built more than 5,400 port facilities for shore power supply and over 280 LNG-powered ships.

Reinforcing energy conservation in public institutions. China has conducted quota management of energy, and published a "leader board" of efficient energy users from public institutions including government bodies, schools and hospitals. China is promoting green building, green office, green travel, green dining room, green IT, and green culture. It has established more than 3,600 energy-saving demonstration units in public institutions.

(2) Rising Green and Low-Carbon Energy Consumption

Expanding the use of electric power in final energy consumption. The "Guiding Opinions on Advancing the Replacement by Electric Energy" has been issued to replace coal and oil with electricity in residential heating, manufacturing, and transport, and to raise the level of electrification of every part of society. In 2019 China used 206.5 billion kWh of electricity to replace less environmentally friendly forms of energy, an increase of 32.6 percent over the previous year.

Strengthening bulk coal management. China has issued the "Implementation Plan for

Upgrading Coal-fired Furnaces for Energy Conservation and Environmental Protection", to improve the efficiency of furnace systems and promote gas-fired furnaces, electric furnaces, and biomass briquette furnaces as per local conditions. China is eliminating small coal-fired boilers in key areas of air pollution control. Restricted zones forbidden to highly polluting fuels have been designated according to air quality control targets.

Advancing clean heating in northern China. China has issued the "Clean Heating Plan for Northern China in Winter (2017 – 2021)" to integrate public wellbeing with environmental control, adopt heating methods suited to local conditions, and promote clean heating. By the end of 2019, the clean heating rate in northern China hit 55 percent, an increase of 21 percentage points over 2016.

Text 4 Building a Clean and Diversified Energy Supply System

Proceeding from its basic national conditions and current stage of development, China gives priority to eco-environmental conservation and pursues green development. It seeks growth while protecting the environment, and believes that a sound eco-environment better facilitates growth. It focuses on supply-side structural reform in the energy sector—giving priority to non-fossil energy, promoting the clean and efficient development and utilization of fossil energy, improving the energy storage, transportation and peak-shaving system, and developing coordinated, complementary, and diverse energy sources in different regions.

1. Prioritizing Non-Fossil Energy

The development and utilization of non-fossil energy is a major element of transitioning to a low-carbon and eco-friendly energy system. China gives priority to non-fossil energy, and strives to substitute low-carbon for high-carbon energy and renewable for fossil energy.

Facilitating the use of solar energy. In line with the principles of driving technological progress, reducing costs, expanding the market and improving the system, China is promoting the use of solar energy in an all-round way. It makes overall planning of geographical layout of solar PV generation bases and market accommodation, with emphasis on both centralized and decentralized power generation. It has implemented a "leader board" incentive to encourage solar PV power generation, and allowed projects to be allocated through market competition, so as to accelerate progress in relevant technologies and reduce costs. As a result, China's solar PV industry has become internationally competitive. The country is improving grid access and other services for decentralized solar PV power generation, and coordinating the development of solar PV power, agriculture, animal husbandry, and desertification control to form a diversified model of solar PV power generation. China is also industrializing solar thermal power generation through demonstration projects, and providing market support for related industrial chains. It has expanded the market for and utilization of solar thermal energy, and introduced centralized hot water projects in industry, commerce and public services to pilot solar heating.

Developing wind power. On the basis of balancing wind power development with power transmission and accommodation, China is taking steps to exploit wind power and building large-scale wind power bases. Based on the principles of overall planning and coordination, and efficient utilization and development of centralized and decentralized wind power both onshore and offshore, it is taking active measures to develop decentralized wind energy in the middle and eastern parts of the country, and offshore wind farms. It gives priority to wind power projects that deliver electricity at affordable prices, and encourages project allocation through market-oriented competition. China also promotes wind power production through large-scale development and utilization of wind power, which helps to boost industry innovation and international competitiveness, and improve the industrial service system.

Developing green hydropower. China considers eco-environmental conservation to be a priority and pursues green development. While protecting the eco-environment and relocating the residents, China develops hydropower in a rational and orderly way, giving equal importance to development and conservation, and emphasizing on both the construction and consequent management of the facilities. Focusing on major rivers in the southwest, China is building large hydropower bases and controlling the construction of small and medium-sized hydropower stations in the basin areas. China seeks the green development of small hydropower stations, and has increased investment in river ecology restoration. It is also improving policies for relocated residents to share the benefits from hydropower projects, thus giving a boost to local economic and social development and helping the relocated population get out of poverty. As in any resource development program, the goals are always set for a better economy, better environment, and better benefits for the people.

Developing safe and structured nuclear power. Nuclear security is the lifeline in developing nuclear power. China attaches equal importance to safety and the orderly development of nuclear power. It has strengthened whole-life management and supervision of nuclear power planning, site selection, design, construction, operation, and decommissioning, and adopted the most advanced technologies and strictest standards for the nuclear power industry. China is improving the multilevel system of regulations and standards on nuclear energy and safety and strengthening relevant emergency plans, legal system, institutions and mechanisms, in its effort to establish a national emergency system that effectively responds to nuclear accidents. China has strengthened nuclear security and nuclear material control, rigorously fulfilling its international obligations towards nuclear security and non-proliferation, and keeping a good nuclear security record. So far, the nuclear power units in operation are generally safe, and there have been no incidents or accidents of level 2 or above on the International Nuclear and Radiological Event Scale.

Developing biomass, geothermal and ocean energy in accordance with local conditions. China is adopting advanced technologies that meet environmental protection standards to generate power by means of urban solid waste incineration, and upgrading biomass power generation to cogeneration of heat and power. It is growing biogas into an industry and transforming methane use in rural areas. In industrializing liquid bio-fuel production by means of non-food biomass, it avoids using crops as raw materials and occupying arable land, strictly controls the expansion of fuel ethanol processing capacity, and focuses on improving the quality of biodiesel products. China is engaged in innovative geothermal power generation, providing urban central heating, and building demonstration zones for efficient production and utilization. It is also reinforcing R&D and pilot demonstrations on harnessing ocean power such as tidal and wave energy.

Increasing the overall utilization rate of renewable energy. China guarantees full acquisition of all renewable energy generated. It has implemented a clean energy accommodation action plan and is adopting various measures to promote the use of clean energy. It is improving the overall planning of the power sector, optimizing the power supply structure and layout, and allowing the market to function as a regulator, to form institutional mechanisms conducive to the use of renewable energy and make the power system more flexible and better at coordinating energy use. China has put in place a mechanism for accommodating power generated from renewable energy, which determines on an annual basis the minimum proportion of renewable energy to be consumed in each province and equivalent administrative unit, and requires suppliers and users to work together to achieve this goal. The country uses the power grid as a platform for optimizing resource allocation. It facilitates optimal interaction and coordination of power source-grid-load-storage, and improves the appraisal and supervision of different sectors in accommodating power generated from renewable energy. Renewable energy use rate has increased significantly: In 2019 the national average consumption rate of wind power was 96 percent, that of solar PV power was 98 percent, and that of water energy in major river basins reached 96 percent.

2. Promoting Clean and Efficient Development and Utilization of Fossil Energy

China coordinates the development and utilization of fossil energy and eco-environmental protection in accordance with its resource endowment and the bearing capacity of natural resources and the environment. It promotes advanced production capacity while phasing out outdated capacity. It also promotes the clean and efficient utilization of coal and the exploration and development of oil and gas, and works to increase reserves and production, so as to be more self-sufficient in oil and gas.

Facilitating the safe, smart and green utilization of coal. China strives to build an intensive, safe, efficient and clean coal industry. It is furthering supply-side structural reform in the industry, improving the coal production capacity replacement policy, speeding up the decommissioning of outdated production facilities, and releasing high-quality capacity in an orderly manner. As a result, the configuration and production capacity of the coal mining sector have seen notable improvement, and large modern coalmines have become the mainstay. From 2016 to 2019, China cut more than 900 million tons of outdated coal production capacity per year on average. It has also increased input in production safety, and improved the mechanism to ensure workplace safety in the long run. Coalmines are becoming highly automated and intelligent by employing more machines and applying information technology, which also make them safer and more efficient. China promotes green mining at large coal bases and facilitates their green transformation by applying coal washing and processing technology, building a circular economy in mining areas and protecting the eco-environment. A number of green mines have been built with improved utilization of various resources in an all-round way. China has taken action to promote the clean and efficient utilization of coal, and increased the quota of coal consumption on power generation. Progress has also been made in coal-to-liquid (CTL) and coal-to-gas (CTG), the precision utilization of low-rank coals, and other industrialized demonstration projects of intensive coal processing.

Promoting the clean and efficient development of thermal power. China has been optimizing coal-fired power and upgrading technology to steadily reduce excess capacity. It has improved the early warning mechanism for risk control in coal-fired power planning and construction, and moved faster to phase out outdated capacity. By the end of 2019, China had phased out more than 100 million kW of outdated coal power capacity, and the ratio of coal-fired power in total power generation had dropped from 65.7 percent in 2012 to 52 percent in 2019. China has taken action to upgrade coal-fired power plants to reduce emissions, and adopted stricter standards for energy efficiency and environmental protection. The efficiency and pollutants control levels of coal-fired power units are on par with world advanced levels. China has also begun to develop natural gas power where appropriate. It encourages adding peak-shaving natural gas power stations to power load centers to improve power security.

Increasing the production of natural gas. In order to increase domestic natural gas supply, China has strengthened basic geological surveying and resource evaluation, and stepped up scientific and technological innovation and industrial support for conventional natural gas production. It is also making breakthroughs in unconventional natural gas exploration and development, such as shale gas and coal-bed gas, and is working on large-scale shale gas exploitation. It is improving relevant policies for the exploitation and utilization of unconventional natural gas. Focusing on the Sichuan, Ordos and Tarim basins, it has built a number of natural gas production bases with an output of more than 10 billion cu m. Since 2017, natural gas output has been increasing by more than 10 billion cu m per year.

Raising the level of oil exploration, development and processing. China has strengthened domestic oil exploration and development, furthering related institutional reforms and promoting scientific and technological R&D and the application of new technologies. It has intensified the exploration and development of low-grade resources, and increased crude oil reserves and production. It has developed advanced oil recovery technologies, increased the recovery ratio of crude oil, and ensured steady output at old oilfields in the east, including the Songliao and Bohai Bay basins. Focusing on the Xinjiang region and the Ordos Basin, it has increased the reserves and production of new oilfields in the west of the country. It has also strengthened offshore oil and gas exploration and development in the Bohai Sea, the East China Sea and the South China Sea, and is advancing deep-sea cooperation with other countries. The output of offshore oilfields was about 40 million tons in 2019. China is also transforming and upgrading its oil refining industry to produce better refined oil products and improved fuel quality, which will reduce exhaust gas pollution of vehicles.

3. Improving the Energy Storage, Transportation and Peak-Shaving System

China coordinates the transportation of various energy resources such as coal, electricity, oil, and gas. It has built interconnected transmission and distribution networks and established a stable and reliable energy storage, transportation and peak-shaving system, to enhance its emergency response.

Strengthening energy transmission and distribution networks. China has been building cross-provincial and cross-regional key energy transmission channels, to connect major energy producing and consumption regions, and promote the complementary and coordinated development between regions. It has improved the capacity of existing railway lines for transporting coal, and seen that more coal is transported by rail with higher efficiency. It has enhanced the connectivity between main natural gas pipelines and provincial pipelines, liquefied natural gas receiving terminals, and gas storages, and is building a unified national network. A natural gas transportation system that is flexible, safe and reliable has taken shape. China has steadily built trans-provincial and trans-regional power transmission channels, and expanded the scope of clean energy allocation in the northwest, north, northeast and southwest. It has improved the main framework of the regional power grid and strengthened internal grid building at the provincial level. It has also carried out flexible HVDC pilot projects, and is working on the Internet of Energy (IoE) and a multilevel power system that is safe, reliable, and of a reasonable size.

Improving energy reserves for emergency response. China has integrated state, corporate, strategic and commercial reserves to achieve higher reserves for oil, natural gas and coal. It has improved the national oil reserve system and accelerated the construction of oil reserve bases. It has set up a multilevel natural gas storage and peak-shaving system, with local governments, gas suppliers, pipeline transportation enterprises and urban gas services fulfilling their respective responsibilities. It has also put in place a coal reserve system with enterprises playing the main part out of their social responsibility and local governments playing a supporting role. China has improved the national emergency mechanism for large-scale power outages, made power supply more reliable, and enhanced its emergency response. It has established a guarantee system for energy transmission and distribution that matches its energy reserve capacity, a standardized system for oil procurement, storage, replacement and use, and a supervisory mechanism for implementation.

Enhancing the energy peak-shaving system. China attaches equal importance to the supply side and the demand side. It strives to increase the peak-shaving capacity with sound market mechanism and strong technological support, so as to use the energy system in an efficient and allround way. It has accelerated the construction of pumped-storage power stations, built natural gas peak-shaving power stations as appropriate, and implemented power flexibility transformation projects in existing coal-fired CHP cogeneration units and coal-fired power generating units, so as to improve the peak-shaving performance of the power system, and promote clean energy accommodation. It is optimizing energy storage, power generation from new energy sources and the operation of the power system, and carrying out electrochemical energy storage and other peak-shaving pilot projects. It has promoted the construction of facilities for natural gas storage and peak shaving, improved the market-oriented mechanism of auxiliary services, and enhanced the peak-shaving capacity of natural gas. China has also improved its policies on electricity and gas prices to guide power and natural gas users to participate in peak shaving and peak shifting, so as to enhance the response on the demand side. It has improved the system for interrupting or adjusting electricity and natural gas load to tap the demand-side potential.

4. Supporting Energy Development in Rural and Poor Areas

China has implemented the strategy of rural revitalization to improve energy security in rural areas, so that the residents can have a better sense of gain, happiness and security.

Improving rural energy infrastructure. Making electricity accessible to all is a basic condition for building a moderately prosperous society in all respects. China implemented a three-year action plan to ensure power access for people without electricity, and had achieved this goal by the end of 2015. China attaches great importance to the renovation and upgrading of the rural power grid and makes great efforts to strengthen the weak links in the process. It has carried out targeted programs for renovating and upgrading power grid in small towns and central villages, connecting motor-pumped wells in rural plain areas to the grid, and supplying poor villages with electricity for industrial and commercial use. Since 2018, it has been focusing on upgrading the power grid in severely impoverished areas and border villages. China has built natural gas branch pipelines and infrastructure to expand the coverage of the pipeline network. To improve energy infrastructure in rural areas, it has built supply outlets for liquefied natural gas, compressed natural gas, and liquefied petroleum gas in areas not covered by natural gas pipelines, and developed renewable energy sources adapted to local conditions.

Carrying out targeted poverty alleviation through energy projects. Energy is a driving force for economic development, and also an important impetus for poverty alleviation. China makes sound plans for the exploitation of energy resources in poor areas, introducing major energy projects in these areas to improve their capability to sustain themselves, thus adding new momentum to the local economy. It has given priority to energy development projects in old revolutionary bases, ethnic minority areas, border areas and poor areas, and built power transmission bases for sending surplus clean electricity to other parts of China, contributing significantly to local economic growth. In developing hydropower, China has followed a sustainable development path by ensuring smooth relocation and resettlement of residents, and by making sure those involved have the means to better themselves, so that the poor share more of the benefits of resource development. China has also increased financial input and policy support for clean energy such as biomass, wind power, solar power, and small hydropower stations in poverty-stricken areas. It has adopted various models integrating solar PV power and agriculture to reduce poverty, and built thousands of "sunshine banks" in poor rural areas.

Using clean energy for heating in rural areas in north China. Winter heating is of great importance in northern China. To ensure that the residents stay warm in winter while reducing air pollution, China has launched clean heating programs in rural areas in accordance with local conditions. In this new scheme enterprises assume the main responsibility and governments provide support to ensure affordable heating for the people. China has been steadily replacing coal with electricity and natural gas for centralized heating, and supported the application of clean biomass fuel, geothermal energy and solar energy in heating, as well as the use of heat pumps. At the end of 2019, the rate of clean-energy heating in the rural areas of north China was about 31 percent, up 21.6 percentage points from 2016. By 2019 about 23 million households in rural areas in northern China had replaced bulk coal with clean energy, including 18 million households in the Beijing-Tianjin-Hebei region, its surrounding areas and the Fenhe-Weihe River Plain.

Notes

(1) supply-side structural reform 供给侧结构性改革：旨在提高经济供给质量和效率的改革战略。

(2) peak-shaving system 调峰系统、调峰体系：通过在低峰期间储存或利用多余能量，减少电力的高峰需求的系统。

（3）geothermal power generation 地热发电：利用地壳内储存的热能发电。

（4）ethanol processing capacity 燃料乙醇加工产能

（5）biodiesel product 生物柴油产品：由可再生资源（如植物油或动物脂肪）制成的燃料。

（6）energy accommodation 能源消纳机制：指一种能源系统中各种能源形式之间的互相接纳和协调机制，确保能源供应的平稳和高效。

（7）coal washing 煤炭洗选：一种通过物理或化学方法从煤炭中去除杂质、减少灰分和硫分含量的过程，提高煤炭品质的方法。

（8）CTL（coal-to-liquid）煤制液体燃料：一种利用煤炭作为原料，通过化学反应将其转化为液体燃料，如合成油和合成柴油等的技术过程。

（9）CTG（coal-to-gas）煤制天然气：一种将煤炭转化为天然气的技术过程，通过气化和化学反应将煤炭转化为可燃性气体。

（10）precision utilization of low-rank coal 低阶煤分质利用：通过使用先进技术和设备，对低质量的煤炭资源进行高效、清洁和经济的利用。

参考译文

第4课 建设多元清洁的能源供应体系

立足基本国情和发展阶段，确立生态优先、绿色发展的导向，坚持在保护中发展、在发展中保护，深化能源供给侧结构性改革，优先发展非化石能源，推进化石能源清洁高效开发利用，健全能源储运调峰体系，促进区域多能互补协调发展。

1. 优先发展非化石能源

开发利用非化石能源是推进能源绿色低碳转型的主要途径。中国把非化石能源放在能源发展优先位置，大力推进低碳能源替代高碳能源、可再生能源替代化石能源。

推动太阳能多元化利用。按照技术进步、成本降低、扩大市场、完善体系的原则，全面推进太阳能多方式、多元化利用。统筹光伏发电的布局与市场消纳，集中式与分布式并举开展光伏发电建设，实施光伏发电"领跑者"计划，采用市场竞争方式配置项目，加快推动光伏发电技术进步和成本降低，光伏产业已成为具有国际竞争力的优势产业。完善光伏发电分布式应用的电网接入等服务机制，推动光伏与农业、养殖、治沙等综合发展，形成多元化光伏发电发展模式。通过示范项目建设推进太阳能热发电产业化发展，为相关产业链的发展提供市场支撑。推动太阳能热利用不断拓展市场领域和利用方式，在工业、商业、公共服务等领域推广集中热水工程，开展太阳能供暖试点。

全面协调推进风电开发。按照统筹规划、集散并举、陆海齐进、有效利用的原则，在做好风电开发与电力送出和市场消纳衔接的前提下，有序推进风电开发利用和大型风电基地建设。积极开发中东部分散风能资源。积极稳妥发展海上风电。优先发展平价风电项目，推行市场化竞争方式配置风电项目。以风电的规模化开发利用促进风电制造产业发展，风电制造产业的创新能力和国际竞争力不断提升，产业服务体系逐步完善。

推进水电绿色发展。坚持生态优先、绿色发展，在做好生态环境保护和移民安置的前提下，科学有序推进水电开发，做到开发与保护并重、建设与管理并重。以西南地区主要河流为重点,有序推进流域大型水电基地建设,合理控制中小水电开发。推进小水电绿色发展,加大对实施河流生态修复的财政投入,促进河流生态健康。完善水电开发移民利益共享政策,坚持水电开发促进地方经济社会发展和移民脱贫致富,努力做到"开发一方资源、发展一方经济、改善一方环境、造福一方百姓"。

安全有序发展核电。中国将核安全作为核电发展的生命线,坚持发展与安全并重,实行安全有序发展核电的方针,加强核电规划、选址、设计、建造、运行和退役等全生命周期管理和监督,坚持采用最先进的技术、最严格的标准发展核电。完善多层次核能、核安全法规标准体系,加强核应急预案和法制、体制、机制建设,形成有效应对核事故的国家核应急能力体系。强化核安保与核材料管制,严格履行核安保与核不扩散国际义务,始终保持着良好的核安保记录。迄今为止在运核电机组总体安全状况良好,未发生国际核事件分级2级及以上的事件或事故。

因地制宜发展生物质能、地热能和海洋能。采用符合环保标准的先进技术发展城镇生活垃圾焚烧发电,推动生物质发电向热电联产转型升级。积极推进生物天然气产业化发展和农村沼气转型升级。坚持不与人争粮、不与粮争地的原则,严格控制燃料乙醇加工产能扩张,重点提升生物柴油产品品质,推进非粮生物液体燃料技术产业化发展。创新地热能开发利用模式,开展地热能城镇集中供暖,建设地热能高效开发利用示范区,有序开展地热能发电。积极推进潮流能、波浪能等海洋能技术研发和示范应用。

全面提升可再生能源利用率。完善可再生能源发电全额保障性收购制度。实施清洁能源消纳行动计划,多措并举促进清洁能源利用。提高电力规划整体协调性,优化电源结构和布局,充分发挥市场调节功能,形成有利于可再生能源利用的体制机制,全面提升电力系统灵活性和调节能力。实行可再生能源电力消纳保障机制,对各省、自治区、直辖市行政区域按年度确定电力消费中可再生能源应达到的最低比重指标,要求电力销售企业和电力用户共同履行可再生能源电力消纳责任。发挥电网优化资源配置平台作用,促进源网荷储互动协调,完善可再生能源电力消纳考核和监管机制。可再生能源电力利用率显著提升,2019年全国平均风电利用率达96%、光伏发电利用率达98%、主要流域水能利用率达96%。

2. 清洁高效开发利用化石能源

根据国内资源禀赋,以资源环境承载力为基础,统筹化石能源开发利用与生态环境保护,有序发展先进产能,加快淘汰落后产能,推进煤炭清洁高效利用,提升油气勘探开发力度,促进增储上产,提高油气自给能力。

推进煤炭安全智能绿色开发利用。努力建设集约、安全、高效、清洁的煤炭工业体系。推进煤炭供给侧结构性改革,完善煤炭产能置换政策,加快淘汰落后产能,有序释放优质产能,煤炭开发布局和产能结构大幅优化,大型现代化煤矿成为煤炭生产主体。2016年至2019年,累计退出煤炭落后产能9亿吨/年以上。加大安全生产投入,健全安全生产长效机制,加快煤矿机械化、自动化、信息化、智能化建设,全面提升煤矿安全生产效率和安全保障水平。推进大型煤炭基地绿色化开采和改造,发展煤炭洗选加工,发展矿区循环经济,加强矿区生态环境治理,建成一批绿色矿山,资源综合利用水平全面提升。实施煤炭清洁高效利用行动,煤炭消费中发电用途占比进一步提升。煤制油气、低阶煤分质利用等煤炭深加工产业化示范取得积极进展。

清洁高效发展火电。坚持清洁高效原则发展火电。推进煤电布局优化和技术升级,积极稳妥化解煤电过剩产能。建立并完善煤电规划建设风险预警机制，严控煤电规划建设，加快淘汰落后产能。截至2019年底，累计淘汰煤电落后产能超过1亿千瓦，煤电装机占总发电装机比重从2012年的65.7%下降至2019年的52%。实施煤电节能减排升级与改造行动，执行更严格能效环保标准。煤电机组发电效率、污染物排放控制达到世界先进水平。合理布局适度发展天然气发电，鼓励在电力负荷中心建设天然气调峰电站，提升电力系统安全保障水平。

提高天然气生产能力。加强基础地质调查和资源评价，加强科技创新、产业扶持，促进常规天然气增产，重点突破页岩气、煤层气等非常规天然气勘探开发，推动页岩气规模化开发，增加国内天然气供应。完善非常规天然气产业政策体系，促进页岩气、煤层气开发利用。以四川盆地、鄂尔多斯盆地、塔里木盆地为重点，建成多个百亿立方米级天然气生产基地。2017年以来，每年新增天然气产量超过100亿立方米。

提升石油勘探开发与加工水平。加强国内勘探开发，深化体制机制改革、促进科技研发和新技术应用，加大低品位资源勘探开发力度，推进原油增储上产。发展先进采油技术，提高原油采收率，稳定松辽盆地、渤海湾盆地等东部老油田产量。以新疆地区、鄂尔多斯盆地等为重点，推进西部新油田增储上产。加强渤海、东海和南海等海域近海油气勘探开发，推进深海对外合作，2019年海上油田产量约4 000万吨。推进炼油行业转型升级。实施成品油质量升级，提升燃油品质，促进减少机动车尾气污染物排放。

3. 加强能源储运调峰体系建设

统筹发展煤电油气多种能源输运方式，构建互联互通输配网络，打造稳定可靠的储运调峰体系，提升应急保障能力。

加强能源输配网络建设。持续加强跨省跨区骨干能源输送通道建设，提升能源主要产地与主要消费区域间通达能力，促进区域优势互补、协调发展。提升既有铁路煤炭运输专线的输送能力，持续提升铁路运输比例和煤炭运输效率。推进天然气主干管道与省级管网、液化天然气接收站、储气库间互联互通，加快建设"全国一张网"，初步形成调度灵活、安全可靠的天然气输运体系。稳步推进跨省跨区输电通道建设，扩大西北、华北、东北和西南等区域清洁能源配置范围。完善区域电网主网架，加强省级区域内部电网建设。开展柔性直流输电示范工程建设，积极建设能源互联网，推动构建规模合理、分层分区、安全可靠的电力系统。

健全能源储备应急体系。建立国家储备与企业储备相结合、战略储备与商业储备并举的能源储备体系，提高石油、天然气和煤炭等储备能力。完善国家石油储备体系，加快石油储备基地建设。建立健全地方政府、供气企业、管输企业、城镇燃气企业各负其责的多层次天然气储气调峰体系。完善以企业社会责任储备为主体、地方政府储备为补充的煤炭储备体系。健全国家大面积停电事件应急机制，全面提升电力供应可靠性和应急保障能力。建立健全与能源储备能力相匹配的输配保障体系，构建规范化的收储、轮换、动用体系，完善决策执行的监管机制。

完善能源调峰体系。坚持供给侧与需求侧并重，完善市场机制，加强技术支撑，增强调峰能力，提升能源系统综合利用效率。加快抽水蓄能电站建设，合理布局天然气调峰电站，实施既有燃煤热电联产机组、燃煤发电机组灵活性改造，改善电力系统调峰性能，促进清洁能源消纳。推动储能与新能源发电、电力系统协调优化运行，开展电化学储能等调峰试点。

推进天然气储气调峰设施建设,完善天然气储气调峰辅助服务市场化机制,提升天然气调峰能力。完善电价、气价政策,引导电力、天然气用户自主参与调峰、错峰,提升需求侧响应能力。健全电力和天然气负荷可中断、可调节管理体系,挖掘需求侧潜力。

4. 支持农村及贫困地区能源发展

落实乡村振兴战略,提高农村生活用能保障水平,让农村居民有更多实实在在的获得感、幸福感、安全感。

加快完善农村能源基础设施。让所有人都能用上电,是全面建成小康社会的基本条件。实施全面解决无电人口问题三年行动计划,2015年底全面解决了无电人口用电问题。中国高度重视农村电网改造升级,着力补齐农村电网发展短板。实施小城镇中心村农网改造升级、平原农村地区机井通电和贫困村通动力电专项工程。2018年起,重点推进深度贫困地区和抵边村寨农网改造升级攻坚。加快天然气支线管网和基础设施建设,扩大管网覆盖范围。在天然气管网未覆盖的地区推进液化天然气、压缩天然气、液化石油气供应网点建设,因地制宜开发利用可再生能源,改善农村供能条件。

精准实施能源扶贫工程。能源不仅是经济发展的动力,也是扶贫的重要支撑。中国合理开发利用贫困地区能源资源,积极推进贫困地区重大能源项目建设,提升贫困地区自身"造血"能力,为贫困地区经济发展增添新动能。在革命老区、民族地区、边疆地区、贫困地区优先布局能源开发项目,建设清洁电力外送基地,为所在地区经济增长作出重要贡献。在水电开发建设中,形成了水库移民"搬得出、稳得住、能致富"的可持续发展模式,让贫困人口更多分享资源开发收益。加强财政投入和政策扶持,支持贫困地区发展生物质能、风能、太阳能、小水电等清洁能源。推行多种形式的光伏与农业融合发展模式,实施光伏扶贫工程,建成了成千上万座遍布贫困农村地区的"阳光银行"。

推进北方农村地区冬季清洁取暖。北方地区冬季清洁取暖关系广大人民群众生活,是重大民生工程、民心工程。以保障北方地区广大群众温暖过冬、减少大气污染为立足点,在北方农村地区因地制宜开展清洁取暖。按照企业为主、政府推动、居民可承受的方针,稳妥推进"煤改气""煤改电",支持利用清洁生物质燃料、地热能、太阳能供暖以及热泵技术应用。截至2019年底,北方农村地区清洁取暖率约31%,比2016年提高21.6个百分点;北方农村地区累计完成散煤替代约2300万户,其中京津冀及周边地区、汾渭平原累计完成散煤清洁化替代约1800万户。

英翻中练习

(1) The Zhangjiakou Renewable Energy Demonstration Zone

In 2015, the state approved the plan for the Zhangjiakou Renewable Energy Demonstration Zone, which proposes "three major innovations" "four major projects" and "five major functional areas". The three major innovations are institutional innovation, business model innovation and technological innovation. The four major projects are large-scale renewable energy development, large-capacity energy storage, intelligent power transmission, and diversified application and demonstration. The five major functional areas are a low-carbon Olympic area, a high-tech incubator, a comprehensive business district, a high-end equipment manufacturing area, and an agricultural recycling demonstration area.

By the end of 2019, the total installed capacity of renewable energy power generation in Zhangjiakou City reached 15 million kW, accounting for more than 70 percent of the total installed power generating capacity in the region. With more than 8 million sq m of floor space being heated by wind power, and 285 million kWh of renewable energy being accommodated by the green data center, renewable energy accounted for 27 percent of regional energy consumption. As of 2019 about 3,000 new energy vehicles had been sold, and a number of hydrogen fuel cell buses had been put into operation. The Zhangbei 500 - kv flexible DC power grid test and demonstration project and the Zhangbei-Xiongan 1,000 - kv ultrahigh voltage AC transmission and transformation project contributed to coordinated development of green energy in the Beijing-Tianjin-Hebei Region.

By 2030, Zhangjiakou's energy supply will be largely based on renewable energy: Eighty percent of its power consumption will be renewable, as will the energy used by all of its public transport, and residential, commercial and public buildings, and its industrial enterprises will all achieve zero carbon emissions.

(2) Achievements in Rural Energy Development and Poverty Alleviation

A new round of upgrading of rural power grids. In 2017, China completed the renovation and upgrading of rural power grids in small towns and central villages, electrified motor-pumped wells, and supplied poor villages with electricity for industrial and commercial use, benefiting 78,000 villages and 160 million rural residents. The country electrified 1.6 million motor-pumped wells, benefiting more than 10,000 townships and 10 million hectares of farmland. It also provided electricity to 33,000 villages. In 2019, China completed a new round of rural grid transformation and upgrading, achieving a supply reliability rate of 99.8 percent and an integrated voltage qualification rate of 97.9 percent. Stable and reliable power supply services have been provided to all rural areas in China.

Universal access to electric power. From 2013 to 2015, the state allocated RMB 24.78 billion to extend power grids to areas without electricity, benefiting some 1.55 million people. It carried out an independent solar PV power supply project, providing electricity to 1.19 million people. By the end of 2015, China had achieved full electricity coverage for its entire population.

Poverty alleviation through solar PV power generation. Poverty alleviation through solar PV power generation is one of the top 10 targeted poverty alleviation projects in China. Since 2014, the state has formulated relevant plans, introduced fiscal, financial and pricing policies, strengthened power grid building and operation services, and promoted various solar PV poverty alleviation projects funded by the government and implemented by enterprises. China has built a total of 26.36 million kW of solar PV power stations for this purpose, benefiting nearly 60,000 poor villages and 4.15 million poor households. The facilities now earn about RMB18 billion from power generation and provide 1.25 million public welfare jobs every year.

Text 5 Leveraging the Role of Innovation as the Primary Driver of Development

China has seized the opportunities presented by the new round of scientific and technological revolution and industrial transformation. In the energy sector, it has implemented a strategy of innovation-driven development to increase its capacity for scientific and technological innovation and address major issues and challenges, such as energy resource constraints, environmental protection, and climate change, through advances in technology.

1. Improving Top Level Design for Energy Policies Relating to Scientific and Technological Innovation

China has made energy a vital part of its innovation-driven development strategy, and given more prominence to innovation in energy science and technology. Modern energy technology that is safe, clean and of high efficiency is a key strategic sector and a national priority in the country's "Outline of Innovation-driven Development Strategy". Accordingly China has drawn blueprints for sci-tech innovation in energy and resources, made strategic plans for scientific and technological development of the resources and energy industry till 2035, and proposed major measures and tasks for innovation in energy science and technology. These are all aimed to enhance the role of scientific and technological innovation in driving and underpinning the energy sector. By making plans for technological innovation in energy and creating the "Innovation Action Plan of Energy Technological Revolution (2016～2030)", China has charted the roadmap and identified its priorities. Through deeper reform, China is establishing an energy science and technology system in which technological innovation is directed by the government and led by the market, and engages the whole of society, with enterprises playing a major role and all stakeholders coordinating with each other. At the same time China has increased investment in scientific and technological innovation in key energy fields and emerging energy industries, stepped up efforts to cultivate professionals in these areas, and endeavored to help all entities involved to improve their capacity for innovation.

2. Creating Diversified Platforms for Technological Innovation in Energy at Various Levels

On the strength of leading enterprises, research institutions and universities, China has created a number of high-standard platforms for technological innovation, and inspired enthusiasm for innovation among all parties involved. Amid efforts to promote scientific and technological advances in energy, China has established more than 40 key national laboratories and a group of national engineering research centers that focus on research into technologies for safe, green and intelligent coal mining, highly efficient use of renewable energy, energy storage, and decentralized energy systems. It has also built more than 80 national energy R&D centers and key national energy laboratories for research in the key areas of coal, oil, natural gas, coal-fired power, nuclear power, renewable energy and energy equipment, all of which cover the vital and frontier areas of energy innovation. Adapting to their own needs and the needs of the industry, large energy enterprises have made continuous efforts to build up their scientific and technological capacity, and have established some influential research institutions in their respective fields. In keeping with the industrial strengths of their regions, local governments have adopted various measures to expand their scientific and technological capacity. Encouraged by the policy of "public entrepreneurship and public innovation", all entities in Chinese society are actively engaged in scientific and technological innovation, and a large number of new energy technology businesses have been established.

3. Promoting Coordinated Scientific and Technological Innovation in Key Realms of the Energy Sector

China has implemented major scientific and technological initiatives and projects to achieve leapfrog development in key energy technologies. Focusing on its strategic industrial goals, China has rolled out a project on oil and gas technology whose emphasis is making breakthroughs in petroleum geology theory and key technologies for high-efficiency exploration and exploitation, as well as finding technology solutions to low-cost, high-efficiency exploitation of unconventional sources of energy, including shale oil, shale gas and gas hydrates. China has launched a project in nuclear power technology to advance research on core technologies of a third-generation pressurized water reactor and a fourth-generation high-temperature gas cooled reactor. The goal is to boost the country's independent innovation in nuclear power technology. In the field of key generic technologies, China has planned for and carried out research into new energy vehicles, smart grid, smart coal mining, clean and efficient use of coal and new energy-saving technology, renewable energy and hydrogen energy, among others. To achieve its major strategic goals, China has given priority to research in basic physics and chemistry concerning clean and efficient use and conversion of energy, in the hope that advances in basic research will lead to breakthroughs in applied technologies.

4. Launching Major Energy Projects to Upgrade Energy Technologies and Equipment

In a global trend of transition to green and low-carbon development in the energy sector, China has accelerated the upgrading of conventional energy technologies and equipment, and is replacing them with new ones at a faster pace. It has redoubled efforts to make independent innovations in emerging energy technologies, and achieved a marked improvement in clean, low-carbon energy technologies. By launching major equipment manufacturing projects and major demonstration projects, China has made breakthroughs in the trials, demonstration, application and popularization of key energy technologies. It has improved the measurement, standard setting, testing, and certification systems of energy equipment, and built up its capacity to research, develop, design and manufacture complete sets of important energy equipment. To achieve secure energy supply, develop clean energy, and encourage the clean, efficient use of fossil fuels, China concentrates on making breakthroughs in key technologies in energy equipment manufacturing, solving bottleneck issues involving materials and accessories, and promoting technological innovation along the whole industrial chain. China has launched major demonstration projects for advanced energy technologies and equipment in such fields as clean and intelligent coal mining, washing and selection, the exploration and exploitation of deep-water and unconventional oil and gas resources, oil and gas storage and transport, clean and efficient coal-fired power generation, advanced nuclear power technologies, power generation from renewable sources, gas turbine, energy storage, advanced power grid, and deep processing of coal.

5. Supporting the Development of New Technologies and New Business Forms and Models

The world now stands at the confluence of a new round of technological revolution and an industrial revolution. New technological breakthroughs have accelerated industrial transformation, giving rise to waves of new business forms and models in the energy sector. China has made strenuous efforts to integrate energy technologies with modern and advanced information, material and manufacturing technologies, and has rolled out the "Internet +" intelligent energy program to explore new models of energy production and consumption. It has stepped up efforts to innovate and upgrade intelligent solar PV power generation, integrate the development of solar PV power generation with agriculture, fishery, animal husbandry and construction, and open new space for the complimentary application of solar PV power generation, creating new models in the utilization of new energy. China has picked up its pace in developing industrial chains in the production, storage, transport and application of green hydrogen, hydrogen-fuel cells, and hydrogen-powered vehicles. It supports the application of energy storage technologies at multiple points in energy production and utilization, and the complementary development of energy storage and renewable energy. By supporting the construction of micro-grids for new energy, China has established regional systems of clean energy supply that integrate power generation, storage and utilization. It promotes new comprehensive energy services and strives for complementary, coordinated and efficient end use of various forms of energy. With pilot and demonstration projects leading the way, a series of new energy technologies and new business forms and models have emerged, triggering a fusion of innovative development in China's energy sector.

Notes

（1）top level design 顶层设计：政府或相关机构制定的整体规划和战略框架，用于指导和协调特定领域政策或计划的实施。

（2）energy resource constraints 能源资源约束、能源资源限制：指能源资源（如化石燃料）的限制或稀缺性可能阻碍可持续发展并需要探索替代能源来源。

（3）innovation action plan of energy technological revolution 能源技术革命创新行动计划：中国制定的促进能源部门技术进步和创新的战略规划或路线图。

（4）coordinated scientific and technological innovation 协同科技创新：指政府、企业、研究机构和社会等不同利益相关方之间的合作努力和协调，以推动特定领域的科技创新。

（5）industrial transformation 产业转型：经济或行业经历结构性变化的过程，通常涉及向新的部门、技术或商业模式的转变。

（6）demonstration project 示范项目

（7）stakeholder 利益相关者

参考译文

第5课　发挥科技创新第一动力作用

抓住全球新一轮科技革命与产业变革的机遇，在能源领域大力实施创新驱动发展战略，增强能源科技创新能力，通过技术进步解决能源资源约束、生态环境保护、应对气候变化等重大问题和挑战。

1. 完善能源科技创新政策顶层设计

中国将能源作为国家创新驱动发展战略的重要组成部分，把能源科技创新摆在更加突出的地位。《国家创新驱动发展战略纲要》将安全清洁高效现代能源技术作为重要战略方向和重点领域。制定能源资源科技创新规划和面向2035年的能源、资源科技发展战略规划，部署了能源科技创新重大举措和重大任务，努力提升科技创新引领和支撑作用。制定能源技术创新规划和《能源技术革命创新行动计划（2016～2030年）》，提出能源技术创新的重点方向和技术路线图。深化能源科技体制改革，形成政府引导、市场主导、企业为主体、社会参与、多方协同的能源技术创新体系。加大重要能源领域和新兴能源产业科技创新投入，加强人才队伍建设，提升各类主体创新能力。

2. 建设多元化多层次能源科技创新平台

依托骨干企业、科研院所和高校，建成一批高水平能源技术创新平台，有效激发了各类主体的创新活力。布局建设40多个国家重点实验室和一批国家工程研究中心，重点围绕煤炭安全绿色智能开采、可再生能源高效利用、储能与分布式能源等技术方向开展相关研究，促进能源科技进步。布局建设80余个国家能源研发中心和国家能源重点实验室，围绕煤炭、石油、天然气、火电、核电、可再生能源、能源装备重点领域和关键环节开展研究，覆盖当前能源技术创新的重点领域和前沿方向。大型能源企业适应自身发展和行业需要，不断加强科技能力建设，形成若干专业领域、有影响力的研究机构。地方政府结合本地产业优势，采取多种方式加强科研能力建设。在"大众创业、万众创新"政策支持下,各类社会主体积极开展科技创新,形成了众多能源科技创新型企业。

3. 开展能源重大领域协同科技创新

实施重大科技项目和工程,实现能源领域关键技术跨越式发展。聚焦国家重大战略产业化目标,实施油气科技重大专项,重点突破油气地质新理论与高效勘探开发关键技术,开展页岩油、页岩气、天然气水合物等非常规资源经济高效开发技术攻关。实施核电科技重大专项,围绕三代压水堆和四代高温气冷堆技术,开展关键核心技术攻关,持续推进核电自主创新。面向重大共性关键技术,部署开展新能源汽车、智能电网技术与装备、煤矿智能化开采技术与装备、煤炭清洁高效利用与新型节能技术、可再生能源与氢能技术等方面研究。面向国家重大战略任务,重点部署能源高效洁净利用与转化的物理化学基础研究,推动以基础研究带动应用技术突破。

4. 依托重大能源工程提升能源技术装备水平

在全球能源绿色低碳转型发展趋势下,加快传统能源技术装备升级换代,加强新兴能源技术装备自主创新,清洁低碳能源技术水平显著提升。依托重大装备制造和重大示范工程,推动关键能源装备技术攻关、试验示范和推广应用。完善能源装备计量、标准、检测和认证体系,提高重大能源装备研发、设计、制造和成套能力。围绕能源安全供应、清洁能源发展和化石能源清洁高效利用三大方向,着力突破能源装备制造关键技术、材料和零部件等瓶颈,推动全产业链技术创新。开展先进能源技术装备的重大能源示范工程建设,提升煤炭清洁智能采掘洗选、深水和非常规油气勘探开发、油气储运和输送、清洁高效燃煤发电、先进核电、可再生能源发电、燃气轮机、储能、先进电网、煤炭深加工等领域装备的技术水平。

5. 支持新技术新模式新业态发展

当前,世界正处在新科技革命和产业革命交汇点,新技术突破加速带动产业变革,促进能源新模式新业态不断涌现。大力推动能源技术与现代信息、材料和先进制造技术深度融合,依托"互联网+"智慧能源建设,探索能源生产和消费新模式。加快智能光伏创新升级,推动光伏发电与农业、渔业、牧业、建筑等融合发展,拓展光伏发电互补应用新空间,形成广泛开发利用新能源的新模式。加速发展绿氢制取、储运和应用等氢能产业链技术装备,促进氢能燃料电池技术链、氢燃料电池汽车产业链发展。支持能源各环节各场景储能应用,着力推进储能与可再生能源互补发展。支持新能源微电网建设,形成发储用一体化局域清洁供能系统。推动综合能源服务新模式,实现终端能源多能互补、协同高效。在试点示范项目引领和带动下,各类能源新技术、新模式、新业态持续涌现,形成能源创新发展的"聚变效应"。

英翻中练习

(1) Breakthroughs in Key Energy Technology and Equipment

Renewable energy technology and equipment. China has mastered key technologies in hydropower, wind and solar energy, and leads the world in the design and manufacturing of complete sets of hydropower generators. New technologies are replacing existing ones at a rapid pace in the whole industrial chain of wind and solar PV electricity, the cost has been falling sharply, and a cohort of internationally competitive companies have emerged. China has also made remarkable progress in biomass energy, geothermal energy and ocean energy technologies.

Power grid technology and equipment. China has fully mastered technologies for ultra-high voltage power transmission and distribution, and conducted demo applications of advanced power grid technologies such as VSC-HVDC and multi-terminal DC transmission. It is making strong headway in the fields of smart grid and large-scale power system control, and leads the world in power transmission and distribution technology and equipment.

Nuclear power technology and equipment. China has mastered the technology to design and build GW-class nuclear power plants with pressurized water reactors. The equipment and the third-generation nuclear power technology developed independently by China are globally advanced. Major progress has been made on the No. 5 unit of Fuqing Nuclear Power Plant, the world's first pilot project using the Hualong-1 design with China's own intellectual property rights. Pilot projects for the pressurized water reactor CAP1400 and high-temperature gas cooled reactor are proceeding smoothly, and breakthroughs have been made in multiple frontier technologies including those associated with fast reactors and advanced small modular reactors.

(2) Breakthroughs in Key Energy Technology and Equipment

Technology and equipment for oil and gas exploration and exploitation. China has developed advanced technologies to explore low-permeability oil fields and heavy oil reservoirs, put into industrial application the technology to safely and efficiently explore mega-sized, ultra-deep gas fields with high sulfur content, and developed a fracturing technology for the formation with ultra-high pressure. It has made important progress in key technologies and equipment for deep sea exploration and exploitation. China has independently developed 3,000 m deep-water semi-submersible drilling platforms, exemplified by HYSY 981, as well as the drilling rigs Bluewhale I and Bluewhale II. All of this marks technological breakthroughs in offshore extraction of natural gas hydrates.

Technology and equipment for clean and efficient coal-fired power generation. China has the capacity to independently develop and build ultra-supercritical coal-fired power generator sets, and brought down coal consumption in power generation to 256 grams of standard coal per kWh. It leads the world in coal-fired power generation in the technologies of air cooling, double reheating, circulating fluidized beds, and ultra-low emissions. China has also built a 100,000-ton-class demo unit for carbon capture, utilization and storage.

Technology and equipment for safe, green and intelligent exploration and use of coal. With its technologies for safe and green extraction of coal among the best in the world, China is transforming coal production to automated, mechanized and intelligent operations. It has developed complete sets of procedures and technologies for deep processing of coal with its own intellectual property rights, such as those for coal gasification and CTL production.

Text 6　Deeper Reform of the Energy System in All Areas

China will fully leverage the decisive role of the market in allocating energy resources, and ensure the government better play its part in this regard. It will extend market-oriented reform in key areas and on vital issues to remove institutional barriers, solve the problem of an incomplete market system, provide strong institutional guarantees for China's energy security and boost the high-quality development of the energy sector.

1. Creating an Energy Market with Effective Competition

China is working hard to cultivate a variety of market entities, break up monopolies, ease market access, and encourage competition. It is building an energy market system that is unified, open, competitive and yet orderly, removing market barriers, and making the allocation of energy resources more efficient and fairer.

Diversifying market entities. China supports a variety of market entities to operate in segments of the energy sector that are not on the entry negative list, in accordance with the law and on equal footing. China has extended systemic reform of oil and gas exploration and exploitation and opened up the market in this regard. It has implemented competitive trading of oil and gas exploration blocks, and adopted a more rigorous exit mechanism for oil- and gas-bearing zones. China encourages qualified enterprises to import crude oil. It has reformed the oil and gas pipeline operation system to separate transport from sales. In an effort to reform electricity distribution, China is opening up electricity distribution and sales to non-government investment in an orderly manner, and is separating power grid enterprises' secondary business from their core business. New market entities are being cultivated in the fields of electricity distribution and sales, energy storage, and comprehensive energy services. Meanwhile China is extending reform of energy SOEs, supporting development of the non-public sector, and conducting active yet prudent mixed-ownership reform in the energy industry to boost the vitality and motivation of energy enterprises.

Building an energy market system that is unified, open, and competitive yet orderly. China has established trading platforms for coal, electricity, petroleum and natural gas to facilitate interaction between demand and supply. A modern coal market system is under construction. Futures trading of thermal coal, coking coal and crude oil and spot trading of natural gas are under way. Restrictions have been lifted on the generation and consumption of electricity by commercial consumers. An electricity market is under way to incorporate medium- and long-term trading, spot trading and other forms of trading of electricity. China is also working to build a unified electricity market across the country and a national carbon emissions trading market.

2. Improving the Market-Based Mechanism for Deciding Energy Prices

Following the principle of "allowing for more competition in electricity generation, sales and consumption while tightening government regulation of power grid, transmission and distribution", China has lifted price control over competitive areas and links. The goal is to allow prices to reflect market demand, and thereby guide the allocation of resources. It has also conducted strict government oversight of the determination of pricing to cover reasonable costs.

Lifting price control over competitive links in an orderly manner. China is steadily fostering a market-based pricing mechanism of commercial electricity generation and distribution, and allowing prices to be decided by electricity users, sellers and producers through market-based modalities. China has extended reform of the price-setting mechanism for on-grid electricity from coal-fired power plants, and introduced a market-based pricing mechanism in which electricity prices may fluctuate above or below the benchmark. Steady progress has been made in determining the price of on-grid electricity from new wind and solar PV power plants through competitive bidding. Relevant parties are encouraged to negotiate on the basis of sharing risks and benefits, and set through market-based modalities the price for trans-provincial or trans-regional transmission of electricity. The pricing mechanism for oil products is being improved. Reform is ongoing in having the market determine gas prices. China has enforced progressive pricing for household consumption of electricity and gas across the nation, ensuring basic living needs are met while encouraging conservation.

Appropriately deciding prices for natural monopoly operations. Allowing recovery of costs plus reasonable profits, China has set appropriate transmission and distribution prices for power grids and gas pipelines. It has analyzed costs and verified prices for electricity transmission and distribution over two regulatory periods. China has also stepped up the price regulation of gas transmission and distribution and analyzed gas costs in order to establish a price regulation system that covers the whole process of gas transmission and distribution.

3. Innovative Management of the Energy Sector and Improved Government Services

Working to become a service-oriented government, the Chinese government has further transformed its functions, streamlined administration, delegated powers, improved regulation and upgraded services. It employs strategic plans and macro-policies on energy, and mobilizes resources for major undertakings. Better oversight and regulation of the energy market will deliver better results and promote fair competition among all market entities. Putting people and lives above everything else, China has remained firm in its commitment to safe production in the energy industry.

Igniting the vitality of market entities. China has extended reform in the energy sector to delegate powers, improve regulation, and upgrade services. This includes reducing approval by the central government for energy projects and delegating the approval power to local authorities for some projects. The requirement for government review and approval has been rescinded for energy projects about which market entities can decide at their own discretion. The number of items of preliminary review has been slashed and the threshold for market access has been lowered, while supervision during and after production has been enhanced and standardized. "Access to electricity" services have been improved. As a result the time, procedures and cost needed for businesses to connect to the power grid have all been cut down. In addition, China has promoted the "internet plus government services" model, and expanded the practice of providing all energy-related government services at one simple window in localities where all relevant authorities have outlets, thereby improving one-stop services.

Guiding the allocation of resources. In addition to other plans, including special plans and action plans, China has drawn up and implemented the "Strategy for Energy Production and Consumption Revolution (2016 – 2030)" for developing the energy sector. These define the overall goals and key tasks, and guide investment in the sector. In order to encourage market entities to appropriately explore and utilize energy resources, China has refined its fiscal, taxation, industrial, financing, and investment policies, implemented a nationwide *ad valorem* tax on crude oil, natural gas and coal, and raised excise tax on oil products. It is building a green finance incentive system to promote new energy vehicles and develop clean energy. China also encourages Renminbi settlement for trading in bulk energy commodities.

Promoting fair competition. China has overhauled the government's regulatory power and responsibilities, and gradually transformed regulation of the electricity sector to comprehensive regulation of the entire energy sector. It has tightened regulation of electricity transaction, distribution and supply, the market order, equitable connection to the power grid, and grid investment, cost and efficiency. China has also reinforced oversight of the opening of oil and gas pipeline facilities to all eligible users, increased information transparency of pipeline operators, and increased the utilization rate of these facilities. Random inspection by randomly selected staff and prompt release of inspection results have been expanded to the whole energy sector. Efforts have been intensified to establish a credit system in the energy industry, created in accordance with law lists of entities that have committed serious acts of bad faith, and to take joint punitive action against such acts, hence increasing the effectiveness of credit regulation. China exercises prudential regulation of new business forms to develop new drivers of growth. It also keeps energy hotlines open to ensure oversight by the public.

Ensuring production safety. China has improved the accountability system for coal mine safety, raised the efficiency of coal mine supervision, regulation and law enforcement, created a standardized management system for coal mine safety, and built up its capacity for disaster prevention and control. As a result coal mine safety has much improved. Enterprises share the main responsibility for power safety, industry regulators share the regulatory responsibility, and local authorities have the overall responsibility for safety in their respective jurisdictions. Oversight and management have been improved to ensure cybersecurity of the electricity system, as well as the safety and quality of electricity construction projects. Safety risks in the supply of electricity are manageable in general, and there has been no instance of extensive blackout. Meanwhile, through stronger safety regulation of the entire oil and gas industrial chain, China has maintained safety in oil and gas production. Thanks to sustained efforts to improve its system and capacity for nuclear safety regulation, China's nuclear power plants and research reactors are generally safe and secure, and the quality of nuclear projects under construction is well controlled as a whole.

4. Improving the Rule of Law in the Energy Sector

Implementation of the rule of law is essential in the energy sector. It stabilizes expectations and creates long-term benefits. China aligns law making with reform and development in the energy industry, and has amended or abolished laws and regulations incompatible with the needs of reform and development in the sector. China adheres to the principle that administrative bodies must fulfill their statutory obligations and must not take any action that is not mandated by law, and sees that the government fully performs its functions in accordance with the law.

Improving the system of energy laws. More laws and administrative regulations concerning the energy sector have been formulated or amended. Supervision and inspection of the enforcement of energy laws and regulations has been intensified. The work to enact, amend or repeal regulations and normative documents in the fields of electricity, coal, oil, natural gas, nuclear power and new energy has been accelerated, in order to incorporate reform results into China's laws, regulations and major policies.

Improving law-based governance of the energy sector. China is working to ensure law-based governance and see that the rule of law materializes in the entire process of making, enforcing, overseeing and managing energy strategies, plans, policies and standards. To raise awareness of the law, it is developing a new paradigm that features interaction and mutual support between the government and enterprises, which will help create an enabling environment across the nation for respecting, studying and observing the law. China has adopted new practices in administrative law enforcement by introducing a nationwide system for disclosing information on administrative law enforcement, a system for recording the entire enforcement process, a system for reviewing the legality of major enforcement decisions, and an accountability system. China will ensure that the channels for applying administrative reconsideration and filing administrative lawsuits remain open, that cases are handled in accordance with laws and regulations that the legitimate rights and interests of administrative counterparts are protected in accordance with the law, and that people can see in every case that justice has been served.

Notes

(1) mixed-ownership reform：混合所有制改革：指在某些国有企业中引入非公有制资本，以提高经济效益和促进改革发展。

(2) ad valorem tax：消费税、以增值税形式征税：是一种根据商品或服务的售价比例来计算和征收税费的一种税收形式。

(3) administrative reconsideration：行政复议：是指当公民或组织对行政行为不服时，向上级行政机关请求重新审查和处理的一种法律程序。

参考译文

第6课　全面深化能源体制改革

充分发挥市场在能源资源配置中的决定性作用，更好发挥政府作用，深化重点领域和关键环节市场化改革，破除妨碍发展的体制机制障碍，着力解决市场体系不完善等问题，为维护国家能源安全、推进能源高质量发展提供制度保障。

1. 构建有效竞争的能源市场

大力培育多元市场主体，打破垄断、放宽准入、鼓励竞争，构建统一开放、竞争有序的能源市场体系，着力清除市场壁垒，提高能源资源配置效率和公平性。

培育多元能源市场主体。支持各类市场主体依法平等进入负面清单以外的能源领域，形成多元市场主体共同参与的格局。深化油气勘查开采体制改革，开放油气勘查开采市场，实行勘查区块竞争出让和更加严格的区块退出机制。支持符合条件的企业进口原油。改革油气管网运营机制，实现管输和销售业务分离。稳步推进售电侧改革，有序向社会资本开放配售电业务，深化电网企业主辅分离。积极培育配售电、储能、综合能源服务等新兴市场主体。深化国有能源企业改革，支持非公有制发展，积极稳妥开展能源领域混合所有制改革，激发企业活力动力。

建设统一开放、竞争有序的能源市场体系。根据不同能源品种特点，搭建煤炭、电力、石油和天然气交易平台，促进供需互动。推动建设现代化煤炭市场体系，发展动力煤、炼焦煤、原油期货交易和天然气现货交易。全面放开经营性电力用户发用电计划，建设中长期交易、现货交易等电能量交易和辅助服务交易相结合的电力市场。积极推进全国统一电力市场和全国碳排放权交易市场建设。

2. 完善主要由市场决定能源价格的机制

按照"管住中间、放开两头"总体思路，稳步放开竞争性领域和竞争性环节价格，促进价格反映市场供求、引导资源配置；严格政府定价成本监审，推进科学合理定价。

有序放开竞争性环节价格。推动分步实现公益性以外的发售电价格由市场形成，电力用户或售电主体可与发电企业通过市场化方式确定交易价格。进一步深化燃煤发电上网电价机制改革，实行"基准价+上下浮动"的市场化价格机制。稳步推进以竞争性招标方式确定新建风电、光伏发电项目上网电价。推动按照"风险共担、利益共享"原则协商或通过市场化方式形成跨省跨区送电价格。完善成品油价格形成机制，推进天然气价格市场化改革。坚持保基本、促节约原则，全面推行居民阶梯电价、阶梯气价制度。

科学核定自然垄断环节价格。按照"准许成本＋合理收益"原则，合理制定电网、天然气管网输配价格。开展两个监管周期输配电定价成本监审和电价核定。强化输配气价格监管，开展成本监审，构建天然气输配领域全环节价格监管体系。

3. 创新能源科学管理和优化服务

进一步转变政府职能，简政放权、放管结合、优化服务，着力打造服务型政府。发挥能源战略规划和宏观政策导向作用，集中力量办大事。强化能源市场监管，提升监管效能，促进各类市场主体公平竞争。坚持人民至上、生命至上理念，牢牢守住能源安全生产底线。

激发市场主体活力。深化能源"放管服"改革，减少中央政府层面能源项目核准，将部分能源项目审批核准权限下放地方，取消可由市场主体自主决策的能源项目审批。减少前置审批事项，降低市场准入门槛，加强和规范事中事后监管。提升"获得电力"服务水平，压减办电时间、环节和成本。推行"互联网＋政务"服务，推进能源政务服务事项"一窗受理""应进必进"，提升"一站式"服务水平。

引导资源配置方向。制定实施《能源生产和消费革命战略（2016—2030）》以及能源发展规划和系列专项规划、行动计划，明确能源发展的总体目标和重点任务，引导社会主体的投资方向。完善能源领域财政、税收、产业和投融资政策，全面实施原油、天然气、煤炭资源税从价计征，提高成品油消费税，引导市场主体合理开发利用能源资源。构建绿色金融正向激励体系，推广新能源汽车，发展清洁能源。支持大宗能源商品贸易人民币计价结算。

促进市场公平竞争。理顺能源监管职责关系，逐步实现电力监管向综合能源监管转型。严格电力交易、调度、供电服务和市场秩序监管，强化电网公平接入、电网投资行为、成本及投资运行效率监管。加强油气管网设施公平开放监管，推进油气管网设施企业信息公开，提高油气管网设施利用率。全面推行"双随机、一公开"监管，提高监管公平公正性。加强能源行业信用体系建设，依法依规建立严重失信主体名单制度，实施失信惩戒，提升信用监管效能。包容审慎监管新兴业态，促进新动能发展壮大。畅通能源监管热线，发挥社会监督作用。

筑牢安全生产底线。健全煤矿安全生产责任体系，提高煤矿安全监管监察执法效能，建设煤矿安全生产标准化管理体系，增强防灾治灾能力，煤矿安全生产形势总体好转。落实电力安全企业主体责任、行业监管责任和属地管理责任，提升电力系统网络安全监督管理，加强电力建设工程施工安全监管和质量监督，电力系统安全风险总体可控，未发生大面积停电事故。加强油气全产业链安全监管，油气安全生产形势保持稳定。持续强化核安全监管体系建设，提高核安全监管能力，核电厂和研究堆总体安全状况良好，在建工程建造质量整体受控。

4. 健全能源法治体系

发挥法治固根本、稳预期、利长远的保障作用，坚持能源立法同改革发展相衔接，及时修改和废止不适应改革发展要求的法律法规；坚持法定职责必须为、法无授权不可为，依法全面履行政府职能。

完善能源法律体系。推进能源领域法律及行政法规制修订工作，加强能源领域法律法规实施监督检查，加快电力、煤炭、石油、天然气、核电、新能源等领域规章规范性文件的"立改废"进程,将改革成果体现在法律法规和重大政策中。

推进能源依法治理。推进法治政府建设,推动将法治贯穿于能源战略、规划、政策、标准的制定、实施和监督管理全过程。构建政企联动、互为支撑的能源普法新格局,形成尊法、学法、守法、用法良好氛围。创新行政执法方式,全面推行行政执法公示制度、行政执法全过程记录制度、重大执法决定法制审核制度,全面落实行政执法责任制。畅通行政复议和行政诉讼渠道,确保案件依法依规办理,依法保护行政相对人合法权益,让人民在每一个案件中切实感受到公平正义。

英翻中练习

(1) Breakthroughs in Market-Oriented Reform of the Electricity Sector

Improving the price regulation system for electricity transmission and distribution. China has established a preliminary regulatory framework based on allowing recovery of costs plus reasonable profits, and changed the profit model of power grid enterprises. This lays the foundations for further market-oriented reform of the electricity sector.

Supporting independent and procedure-based operation of electricity trading agencies. China has established two regional trading agencies, in Beijing and Guangzhou, and 33 agencies at the provincial level. It is also working to improve the management of trading agencies by transforming them into joint stock companies.

Opening up electricity distribution and sales business. China encourages the participation of non-government investment in electricity distribution, and encourages qualified enterprises to engage in electricity sales. This will give consumers more choice. By the end of 2019 the country had launched 380 pilot reform projects to expand the business of electricity distribution, and the number of companies selling power that were registered with electricity trading agencies had risen to proximately 4,500.

Developing the electricity market. China has lifted restrictions on electricity generation and consumption in an orderly manner, promoted mid- and long-term trading of electricity, and rolled out a pilot program for spot trading in eight regions. It is developing the market for supporting services among five trans-regional and 27 provincial power grids. In 2019 market-based transactions in electricity totaled 2.71 trillion kWh, accounting for 37.5 percent of China's power consumption for that year.

(2) Better Electricity Access for Businesses

Increasing electricity access to improve the business environment is key to giving market entities and the general public a stronger sense of fulfillment and greater satisfaction. China has promoted a service to connect small and micro businesses applying for low capacity to the power grid without visits to government offices, government approval or investment by the applicants. At the end of 2019 this service had been made available in all municipalities directly under the central government, provincial capitals, and capitals of autonomous regions, and the procedure could be completed within 30 working days. World Bank reports show that from 2017 to 2019, the average number of steps required for a business to connect to the grid in China was cut from 5.5 to 2, resulting in marked reduction in the amount of time and money they spent in the process, and China's ranking on the Getting Electricity Indicator rose all the way from No. 98 to No. 12.

Text 7 Strengthening International Energy Cooperation Across the Board

China bases international cooperation on the principle of mutual benefit and win-win results while embracing the concept of green development. It is endeavoring to ensure energy security in an open environment, open its energy sector wider to the world, promote high-quality Belt and Road cooperation, actively engage in global energy governance, guide global cooperation in climate change, and build a global community of shared future.

1. Opening the Energy Sector Further to the World

China is committed to a stable global energy market and is opening its energy sector wider to the world. It has greatly eased market access for foreign investment, and has built a market-based international business environment that respects the rule of law to facilitate free trade and investment. It has adopted pre-establishment national treatment plus a negative list, reducing restrictions on access to the energy sector for foreign investment. It has lifted the restrictions for foreign investment to enter the sectors of coal, oil, gas, electric power (excluding nuclear power), and new energy. It is promoting the energy industry in pilot free trade zones such as Guangdong, Hubei, Chongqing and Hainan, and supports further opening up of the entire oil and gas industry in the China (Zhejiang) Pilot Free Trade Zone. International energy companies such as ExxonMobil, GE, BP, EDF and SIEMENS are steadily expanding investment in China. Major foreign investment projects such as Tesla's Shanghai plant are well under way. Foreign-funded gas stations are spreading.

2. Promoting Energy Cooperation Among BRI Countries

China follows the principles of extensive consultation, joint contribution and shared benefits, and pursues open, green and clean governance in its energy cooperation with BRI countries towards high-standard, people-centered and sustainable goals. It attempts to bring benefits to more countries and their people while maintaining its own development trajectory, and to create conditions favorable to further common development.

Pragmatic and mutually beneficial energy cooperation. China engages in extensive cooperation with over 100 countries and regions around the world in terms of energy trade, investment, industrial capacity, equipment, technology, and standard setting. The high standards of Chinese enterprises are much sought after by partner countries for their energy projects, which help to turn local resource advantages into development strengths. They will also drive technical progress in these countries, create more jobs, stimulate the economy, and improve people's lives. In this way China and its BRI partners will grow together by leveraging and incorporating their respective strengths. China builds cooperation with countries and large transnational corporations in the field of clean energy through third-party markets, to create an energy cooperation framework which is open, transparent, inclusive, and mutually beneficial. In 2019, China established Belt and Road energy partnerships with 30 countries.

A silk road with green energy. China is the largest renewable energy market and the largest clean energy equipment manufacturer in the world. It is actively working towards green and low-carbon global energy transition by engaging in extensive cooperation in renewable energy. Its efforts can be seen in cooperation projects such as the Kaleta hydropower project in Guinea, the Kaposvar PV power station project in Hungary, the Mozura wind park project in Montenegro, Noor Energy 1—the CSP + PV solar power project in Dubai of the UAE, the Karot hydropower project and the first phase of the solar PV power project in the Quaid-e-Azam Solar Park in Pakistan. The wide application of renewable energy technologies in the Chinese market is helping to reduce the cost of renewable energy across the globe and accelerate the green transition process.

Greater energy infrastructure connectivity. China is promoting transnational and cross-regional energy infrastructure connectivity, creating conditions for complementary cooperation and reciprocal trade in energy resources. A batch of landmark energy projects such as the China-Russia, China-Central Asia and China-Myanmar oil and gas pipelines have been completed and brought into operation. China has now connected its grid with the power grids of seven neighboring countries, giving a strong boost to energy infrastructure connectivity and realizing optimal allocation of energy resources on a larger scale, which facilitates economic cooperation within the region.

Wider global energy access. China actively implements the UN sustainable development goal of ensuring "access to affordable, reliable, sustainable and modern energy for all". It also takes an active part in global cooperation on expanding energy access. To improve energy access in partner countries and benefit ordinary people, China has employed multiple financing methods to develop electric power projects using grid-connected, microgrid, or off-grid solar systems according to local conditions, and donated clean cooking stoves to regions still using traditional cooking fuels.

3. Actively Participating in Global Energy Governance

As a staunch supporter of multilateralism, China builds bilateral and multilateral energy cooperation based on mutual benefit and win-win results. It supports the role of the International Energy Agency (IEA) and relevant cooperation mechanisms in global energy governance, promotes global energy market stability and supply security, and the green energy transition within the framework of international multilateral cooperation, and contributes ideas and solutions to the sustainable development of global energy.

Engagement in multilateral energy governance. China is an active participant in international energy cooperation under multilateral mechanisms such as the UN, G20, APEC and BRICS. It is making positive progress in joint research, releasing reports and founding agencies. China has set up intergovernmental energy cooperation mechanisms with over 90 countries and regions, and established ties with over 30 international organizations and multilateral mechanisms in the energy sector. Since 2012, China has become a member state of the International Renewable Energy Agency (IRENA), an observer country to the Energy Charter Treaty, and an affiliate of the IEA.

A facilitator in regional energy cooperation. China has built regional energy cooperation platforms with ASEAN, the League of Arab States, African Union, and Central and Eastern Europe, and organized forums on clean energy at the East Asia Summit. It has also facilitated capacity building and cooperation on technological innovation and provided training for 18 countries in clean energy use and energy efficiency.

4. Joining Forces to Tackle Global Climate Change

Embracing the vision of a global community of shared future, China works together with other countries to address global climate change and promote the transition to green and low-carbon energy.

Strengthening international cooperation on climate change. With support from the UN, World Bank, Global Environment Facility, Asian Development Bank, and countries such as Germany, China is focusing on green and low-carbon energy transition and developing extensive and sustainable bilateral and multilateral cooperation with other countries in exploiting renewable energy and showcasing pilot low-carbon cities through experience sharing, technical exchanges, and project dovetailing.

Supporting capacity building in developing countries to address climate change. China is committed to deeper South-South climate cooperation. It provides support to the least developed countries, small island countries, African countries and other developing countries in their response to climate change. Since 2016, China has set up 10 pilot low-carbon industrial parks, launched 100 mitigation and adaptation programs, and provided 1,000 training opportunities on climate change cooperation in developing countries to help them develop clean and low-carbon energy and jointly address global climate change.

5. China's Proposals for Developing Synergy on Sustainable Global Energy Development

Humanity has entered an era of connectivity when maintaining energy security and addressing global climate change have become major challenges confronting the whole world. The ongoing Covid-19 pandemic highlights all the more the interdependent interests of all countries and the interconnection of all peoples. China proposes that the international community should work together on the sustainable development of global energy, address the challenges of climate change, and build a cleaner and more beautiful world.

Jointly promoting the transition to green and low-carbon energy to build a cleaner and more beautiful world. It requires the joint effort of all countries to address the challenge of climate change and improve the global eco-environment. All countries should choose the green development path, adopt green, low-carbon and sustainable working practices and lifestyles, promote energy transition, and address problems relating to energy. We should join forces to tackle global climate change and make our contribution to building a cleaner and more beautiful world.

Jointly consolidating multilateral energy cooperation to accelerate the green economic recovery and growth. We should improve international governance and maintain an open, inclusive, balanced and reciprocal multilateral framework for international energy cooperation. We should expand communication and pragmatic cooperation in the energy sector to promote economic recovery and integrated development. We should strengthen transnational and cross-regional innovation on clean-energy and low-carbon technologies, and cooperation on technology standards, to promote energy technology transfer and rollout and improve international IPR protection.

Jointly facilitating international investment in energy trading to protect global market stability. We should eliminate energy trade and investment barriers, facilitate trade and investment, cooperate on energy resources and industrial capacity as well as infrastructure, improve connectivity, and promote efficient resource allocation and greater market integration. We should embrace the principles of extensive consultation, joint contribution and shared benefits, seek the greatest common ground to promote the sustainable development of global energy, and jointly maintain global energy security.

Jointly improving energy access in underdeveloped areas to address energy poverty. We should join forces to realize the sustainable goal in the energy sector, and ensure access to basic energy services such as electricity for people in need in underdeveloped countries and regions. We should help underdeveloped countries and regions to popularize advanced green energy technologies, train energy professionals and improve energy services to integrate the efforts on green energy development and the elimination of energy poverty.

Notes

(1) market access 市场准入：允许个人或公司进入并参与特定市场或行业。
(2) negative list 负面清单：指规定外国投资受限或禁止的领域或行业的清单。
(3) pilot free trade zone 自由贸易试验区

参考译文

第 7 课　全方位加强能源国际合作

中国践行绿色发展理念，遵循互利共赢原则开展国际合作，努力实现开放条件下能源安全，扩大能源领域对外开放，推动高质量共建"一带一路"，积极参与全球能源治理，引导应对气候变化国际合作，推动构建人类命运共同体。

1. 持续深化能源领域对外开放

中国坚定不移维护全球能源市场稳定，扩大能源领域对外开放。大幅度放宽外商投资准入，打造市场化法治化国际化营商环境，促进贸易和投资自由化便利化。全面实行准入前国民待遇加负面清单管理制度，能源领域外商投资准入限制持续减少。全面取消煤炭、油气、电力（除核电外）、新能源等领域外资准入限制。推动广东、湖北、重庆、海南等自由贸易试验区能源产业发展，支持浙江自由贸易试验区油气全产业链开放发展。埃克森美孚、通用电气、碧辟、法国电力、西门子等国际能源公司在中国投资规模稳步增加，上海特斯拉电动汽车等重大外资项目相继在中国落地，外资加油站数量快速增长。

2. 着力推进共建"一带一路"能源合作

中国秉持共商共建共享原则，坚持开放、绿色、廉洁理念，努力实现高标准、惠民生、可持续的目标，同各国在共建"一带一路"框架下加强能源合作，在实现自身发展的同时更多惠及其他国家和人民，为推动共同发展创造有利条件。

推动互利共赢的能源务实合作。中国与全球 100 多个国家、地区开展广泛的能源贸易、投资、产能、装备、技术、标准等领域合作。中国企业高标准建设适应合作国迫切需求的能源项目，帮助当地把资源优势转化为发展优势，促进当地技术进步、就业扩大、经济增长和民生改善，实现优势互补、共同发展。通过第三方市场合作，与一些国家和大型跨国公司开展清洁能源领域合作，推动形成开放透明、普惠共享、互利共赢的能源合作格局。2019 年，中国等 30 个国家共同建立了"一带一路"能源合作伙伴关系。

建设绿色丝绸之路。中国是全球最大的可再生能源市场，也是全球最大的清洁能源设备制造国。积极推动全球能源绿色低碳转型，广泛开展可再生能源合作，如几内亚卡雷塔水电项目、匈牙利考波什堡光伏电站项目、黑山莫茹拉风电项目、阿联酋迪拜光热光伏混合发电项目、巴基斯坦卡洛特水电站和真纳光伏园一期光伏项目等。可再生能源技术在中国市场的广泛应用，促进了全世界范围可再生能源成本的下降，加速了全球能源转型进程。

加强能源基础设施互联互通。积极推动跨国、跨区域能源基础设施联通，为能源资源互补协作和互惠贸易创造条件。中俄、中国—中亚、中缅油气管道等一批标志性的能源重大项目建成投运，中国与周边 7 个国家实现电力联网，能源基础设施互联互通水平显著提升，在更大范围内促进能源资源优化配置，促进区域国家经济合作。

提高全球能源可及性。积极推动"确保人人获得负担得起的、可靠和可持续的现代能源"可持续发展目标的国内落实，积极参与能源可及性国际合作，采用多种融资模式为无电地区因地制宜开发并网、微网和离网电力项目，为使用传统炊事燃料的地区捐赠清洁炉灶，提高合作国能源普及水平，惠及当地民生。

3. 积极参与全球能源治理

中国坚定支持多边主义，按照互利共赢原则开展双多边能源合作，积极支持国际能源组织和合作机制在全球能源治理中发挥作用，在国际多边合作框架下积极推动全球能源市场稳定与供应安全、能源绿色转型发展，为促进全球能源可持续发展贡献中国智慧、中国力量。

融入多边能源治理。积极参与联合国、二十国集团、亚太经合组织、金砖国家等多边机制下的能源国际合作，在联合研究发布报告、成立机构等方面取得积极进展。中国与90多个国家和地区建立了政府间能源合作机制，与30多个能源领域国际组织和多边机制建立了合作关系。2012年以来，中国先后成为国际可再生能源署成员国、国际能源宪章签约观察国、国际能源署联盟国等。

倡导区域能源合作。搭建中国与东盟、阿盟、非盟、中东欧等区域能源合作平台，建立东亚峰会清洁能源论坛，中国推动能力建设与技术创新合作，为18个国家提供了清洁能源利用、能效等领域的培训。

4. 携手应对全球气候变化

中国秉持人类命运共同体理念，与其他国家团结合作、共同应对全球气候变化，积极推动能源绿色低碳转型。

加强应对气候变化国际合作。在联合国、世界银行、全球环境基金、亚洲开发银行等机构和德国等国家支持下，中国着眼能源绿色低碳转型，通过经验分享、技术交流、项目对接等方式，同相关国家在可再生能源开发利用、低碳城市示范等领域开展广泛而持续的双多边合作。

支持发展中国家提升应对气候变化能力。深化气候变化领域南南合作，支持最不发达国家、小岛屿国家、非洲国家和其他发展中国家应对气候变化挑战。从2016年起，中国在发展中国家启动10个低碳示范区、100个减缓和适应气候变化项目和1 000个应对气候变化培训名额的合作项目，帮助发展中国家能源清洁低碳发展，共同应对全球气候变化。

5. 共同促进全球能源可持续发展的中国主张

人类已进入互联互通的时代，维护能源安全、应对全球气候变化已成为全世界面临的重大挑战。当前持续蔓延的新冠肺炎疫情，更加凸显各国利益休戚相关、命运紧密相连。中国倡议国际社会共同努力，促进全球能源可持续发展，应对气候变化挑战，建设清洁美丽世界。

协同推进能源绿色低碳转型，促进清洁美丽世界建设。应对气候变化挑战，改善全球生态环境，需要各国的共同努力。各国应选择绿色发展道路，采取绿色低碳循环可持续的生产生活方式，推动能源转型，协同应对和解决能源发展中的问题，携手应对全球气候变化，为建设清洁美丽世界作出积极贡献。

协同巩固能源领域多边合作，加速经济绿色复苏增长。完善国际能源治理机制，维护开放、包容、普惠、平衡、共赢的多边国际能源合作格局。深化能源领域对话沟通与务实合作，推动经济复苏和融合发展。加强跨国、跨地区能源清洁低碳技术创新和标准合作，促进能源技术转移和推广普及，完善国际协同的知识产权保护。

协同畅通国际能源贸易投资，维护全球能源市场稳定。消除能源贸易和投资壁垒，促进贸易投资便利化，开展能源资源和产能合作，深化能源基础设施合作，提升互联互通水平，促进资源高效配置和市场深度融合。秉持共商共建共享原则，积极寻求发展利益最大公约数，促进全球能源可持续发展，共同维护全球能源安全。

协同促进欠发达地区能源可及性，努力解决能源贫困问题。共同推动实现能源领域可持续发展目标，支持欠发达国家和地区缺乏现代能源供应的人口获得电力等基本的能源服务。帮助欠发达国家和地区推广应用先进绿色能源技术,培训能源专业人才,完善能源服务体系,形成绿色能源开发与消除能源贫困相融合的新模式。

> 英翻中练习

(1) Foreign Investment Is Given Wider Access to the Energy Sector

In 2017, the "Catalogue for the Guidance of Industries for Foreign Investment" was revised. For the first time this gave a negative list for foreign investment access to be adopted across the whole country. The Special Administrative Measures (Negative List) for Foreign Investment Access under the Catalogue were separately released in 2018. Equal treatment has since been given to domestic and foreign investment in all sectors outside the negative list.

The Negative List for foreign investment issued in 2018 removed the following access restrictions:

- Construction and operation of power grids (with Chinese party as the controlling shareholder)
- Exploration and exploitation of special and rare kinds of coal (with Chinese party as the controlling shareholder)
- Manufacturing of complete vehicles using new energy (the proportion of Chinese shares shall not be less than 50%)
- Construction and operation of gas stations (in the case of the same foreign investors selling product oil of different varieties and brands from multiple suppliers through more than 30 chain gas stations, the Chinese parties shall be the controlling shareholders)

The Negative List issued in 2019 removed the following access restrictions:

- Exploration and exploitation of oil and natural gas (including coal-bed gas and excluding oil shale, oil sand, shale gas and so on) (limited to Sino-foreign equity or contractual joint ventures)
- Construction and operation of pipeline networks for gas and heat supply in cities with a population of more than 500,000 (with Chinese party as the controlling shareholder)

(2) China's Efforts to Improve the Global Energy Governance System

Within the framework of international multilateral cooperation, China actively promotes the formulation and implementation of initiatives on global energy market stability and supply security, the transition to green and low-carbon energy, greater access to energy, and higher energy efficiency.

- Advocate of a global energy network to meet global power demand with clean and green alternatives

- Facilitator of the G20 Energy Efficiency Leading Programme, the Enhancing Energy Access in Asia and the Pacific: Key Challenges and G20 Voluntary Collaboration Action Plan, and the G20 Voluntary Action Plan on Renewable Energy
- Co-host of the International Forum on Energy Transitions in cooperation with the IRENA and other international organizations
- Proponent of the Shanghai Cooperation Organization Energy Club
- Host of the APEC Sustainable Energy Center in China
- Proponent of the BRICS Energy Research Cooperation Platform
- Founding Member of the IEA's new Energy Efficiency Hub

附录　能源、电力、工程管理英汉术语表

1. 能源

(air)blower	鼓风机
a resource-conserving and environmentally friendly society	资源节约型、环境友好型社会
air preheater	空气预热器
alleviation	缓和
alternative/renewable energy	可替代/可再生能源
alternative fuel	可替代燃料
atomic reactor	原子反应堆
balanced ecological system	生态平衡系统
battery	电池
biodegradation	生物降解
bio-fuel	生物燃料
biogas	生物气，沼气
biogas	沼气
biological weapons	生物武器
biomass energy	生物质能
boiler	锅炉
carbon emitter	温室气体排放国
carbon footprint	炭排放量
carbon monoxide	一氧化碳
carbon-neutral cities	零碳城市
charcoal	木炭
chemical oxygen demand	化学需氧量

续　表

clean and renewable energy sources	清洁、可再生能源
clean energy revolution	清洁能源革命
coal	煤炭
coal conveyor	运煤机
coal ash handling system	煤灰处理系统
coal hopper	煤漏斗
coal pulverizer	碎煤机
coal-fired power plants	燃煤电厂
combustion	燃烧
combustion system	燃烧系统
compressor	压缩机
condenser	冷凝器
conducting fluid	导电流体
conservation	保护
consumption	消费
convection	对流
conventional	常规的
corrosion	腐蚀
cycle	周期
damper	减震器
deaerator	除氧器
demineralization	除盐
depletion	耗竭
ebb	退潮
elasticity of energy consumption	能源消费弹性系数
electrical system	电气系统
electricity	电
energy conservation and emission reduction technologies	节能减排技术
energy consumption	能源消耗
energy conversion	能源转换
energy efficiency	能源效率

续 表

energy policy	能源政策
energy production	能源生产
energy utilization patterns	能源利用方式
energy-efficiency standards rate	能效标准率
energy-efficient technology	高能效技术
engine	发动机
ethanol	乙醇，酒精（新的汽车燃料）
evaporation	蒸发
extract	榨取
feed-water heater	给水加热器
fire	火
fission	分裂
flue gas	烟气
fossil fuel	化石燃料
fuel alcohol	燃料酒精
furnace	熔炉
fusion	熔合物
gas liquefaction	气体液化
gasoline	汽油
generator	发电机
generator motor	发电电动机
geothermal	地热的
geothermal energy	地热能
geothermal power plant	地热发电站
GHG emission reduction	温室气体减排
green car	新能源汽车
high pressure steam turbine	高压气轮机
high voltage electrical distribution	高压配电
hopper	漏斗
hydraulic power	水力
hydro energy	水能

续 表

hydroelectric	水电的
hydroelectric generation	水力发电
hydrogen	氢
inexhaustible	无穷无尽的,永不枯竭的
infrared rays/ultrared rays	红外线
inorganic substances	无机物质
intermediate pressure steam turbine	中压汽轮机
International Solid Waste Association (ISWA)	国际固体废物协会
irrigation	灌溉
Kyoto Protocol	京都议定书
leakage	渗漏
liquefied gas	液化气
magnetic field	磁场
magneto-hydrodynamic power generator	磁流体发电机
mechanical	力学的
moisture content	含水量
motor	马达
natural bitumen	天然沥青
natural gas	天然气
natural gas exploration	天然气勘探
New Energy Security Strategy featuring *Four Reforms and One Cooperation*	"四个革命,一个合作"能源安全新战略
new-energy model city	新能源示范城
nitrogen fixation	固氮
non-renewable energy	一次能源
nuclear power	核能,原子能
ocean thermal energy conversion	海洋热能转换系统
offshore oil drilling	近海石油钻探
oil shales	油页岩
oil well	油井
passive	无源的

续 表

petroleum	石油
photovoltaic power generation	光伏发电
pollution control in major river valleys and regions	重点流域的污染防治
power	电源
powerhouse complex	厂房
pure daily-regulated pumped storage power plant	日调节纯抽水蓄能电站
radioactive	放射性的
reactor	反应堆
refractory	耐火材料
respiratory ills	呼吸道疾病
reversible	可翻转的
rural drinking water	农村饮用水
safe drinking water	安全饮用水
separator	离析器
sewage treatment plant	污水处理厂
silicon cell	硅电池
smart grid	智能电网
solar energy	太阳能
specific heat (capacity)	比热容
steam	蒸汽
steam control valve	蒸汽调节阀
steam water system	汽水系统
storage power plant	蓄能电站
sulfur dioxide	二氧化硫
sulphur compound	硫化合物
superheated vapor	过热蒸汽
superheater	过热装置
tar sands	沥青沙,焦油沙
targets for saving energy and reducing emissions	节能减排目标
thermal energy	热能
thermal sea power	海洋热能

续 表

tidal energy	潮汐能
tidal power station	潮汐发电站
to close down backward production facilities	淘汰落后生产设施
to develop advanced production facilities	加强先进生产设施建设
trail	跟踪，追踪
transformers	变压器
transmission lines	输电线
trough	水槽
turbine	涡轮
ultraviolet rays	紫外线
urban sewage treatment capacity	城市污水处理能力
ventilation device	通风装置
viscosity	粘度
volatile matter	挥发物
waste recovery	废物回收
water conveyance system	输水系统
water power	水力
wave energy	波浪能
windmill	风车
wood-burning	烧木柴的

2. 电力

a half sine wave	半个正弦波
abnormal discharge	异常放电
abrupt change	陡变
accelerant	催化剂
acceptor circuit	带通电路
accordant connection	匹配连接
action-current	作用电流
active power	有功功率

actuating [driving] system	传动系统
after-condenser	二次冷凝器
after-potential	后电势
air-blast circuit breaker	空气吹弧断路器
air-cored current limiting series reactor	空心的串联限流电抗器
alternating current voltage	交流电压
ammeter	电流计
ammonia sniffing	氨气检漏
Ampere(A)	安(培)
ampere-turns	安匝
ample [high] power	大功率
amplifier-rectifier	放大整流器
angular rated power frequency	功率额定的角频率
ant-apex	背点
anti-checking iron	扒钉
apparatus	电器
apparent power	表观功率
arc model parameter	电弧模型参数
arc spot welding	电弧点焊
arc voltage waveform	电弧电压波形
arc-extinguishing medium/arc-interrupting medium	灭弧介质
arcing withstand time	耐弧时间
arrester protective level	避雷器保护水平
assembly drawing	安装图
assembly of distribution apparatus	成套配电装置
auxiliary opening circuit/second tripping circuit	辅助分闸回路
back-pressure turbine	背压式汽轮机
back-to-back capacitor bank	背对背电容器组
backup	备份
back-up circuit breaker/protection circuit breaker	保护断路器
back-up switch/protection switch	保护开关

baking	焙烧
bar(gauge pressure)	巴（表压）
be accurate within	精度达
be held in closed position	保持在合闸位置
belted cable	带绝缘电缆
billet	棒料
blank panel	备用面板
blocking element	闭锁元件
blow-out coil	吹弧线圈
bourdon tube pressure sensor	布尔登管式传感元件
breakdown of insulation	绝缘击穿
breakdown voltage	击穿电压
breaker	断路器
bubble-discharge	气隙放电
bulk optical current transformer	大型光电电流互感器
bursting disks	爆破片
busbar	母线
busbar surge impedance	母线波阻抗
butterfly nut/thumb nut	蝶形螺母
bypass system	旁路系统
cable connection enclosure	电缆连接外壳
cable installation	电缆安装
camber	侧向弯度
capacitance potential divider	电容分压器
capacitor	电容器
capacitor bank-breaking capacity	电容器组开断能力
carry low over-current	承载低的过电流
charge dissipation/accumulation	电荷耗散/累积
charge monitoring relay	充电监视继电器
charge simulation method	模拟电荷法
charging area/charge make-up area	备料场

续　表

charging current interruption test	充电电流开断试验
circuit-breaker not for auto-reclosing	非自动重合闸断路器
circulating high-frequency current	循环高频电流
clad conductor	包覆导体
closed circuit	闭合电路
closed-loop control	闭环控制
closing circuit's d.c power supply/dc supply for closing circuit	合闸回路直流电源
coated wire	被覆线
coating layer/coat/coating	被覆层
coaxial shunt resistor	轴向并联电阻器
co-generation power plant	废热发电厂
color index	比色指数
colorimetric analysis	比色分析
combined overvoltage	综合过电压
comparative reference	比较基准
conducting particle	导电粒子
conduction current	传导电流
conductivity	导电性
connect in parallel to/in parallel/parallel connection	并联
constant arc-voltage	恒定的电弧电压
constant value of the arc voltage	电弧电压稳定值
contact assembly	触点组件
contact bouncing	触点回跳
contact chatter	触点抖动
contact friction forces	触头摩擦力
contact of short-circuit lock-out	短路闭锁接点
contact over-travel/penetrating distance of contact	触头超程
contact voltage/touch voltage	接触电压
contact wedge	电阻片
continuous (power-frequency) voltage	持续(工频)电压
continuous operating voltage	持续运行电压

control switch for closing lockout	合闸闭锁控制开关
control valve/application valve/controlled valve	控制阀
conventional current transformer	常规电流互感器
converter	变流器
converter transformer	变流变压器
corona initial line	电晕起始线
creepage ratio distance	爬电比距
critical damping resistance	临界阻尼电阻
cross-load	横向负荷
cup-shaped axial magnetic structure	杯状纵磁结构
current	电流
current before power-frequency current-zero	工频电流零点前电流
current path	电流通路
current rating	电流额定值
current source driving voltage/current source voltage	电流源电压
current transformer secondaries	电流互感器二次绕组
current-carrying capacity	承受电流的能力
D.C. power supply	直流电源
Da Ya Bay Nuclear Power Plant	大亚湾核电站
dam site	坝址
damping time constant of discharging current	放电电流衰减时间常数
de-energized	不带电的
density relay	密度继电器
detent	掣子
deteriorating effects due to arc-energy	电弧能量产生的烧损反应
Development and Reform Commission	发改委
diagnostic model of power transformer fault	电力变压器故障诊断模型
dielectric constant/permittivity	介电常数
diesel-electric locomotive	柴油电动机车
differential relay	差动继电器
dimensional tolerance	尺寸公差

续 表

direct current voltage	直流电压
disconnector with a vertical isolating distance	垂直隔离断口的隔离开关
dispatching automation	调度自动化
disruptive discharge/sparkover	击穿放电
distortion current	畸变电流
distributed control system	分散控制系统
distribution transformer	配电变压器
divided support disconnector	底座分离式隔离开关
draft	拔模斜度
restoring circuit	恢复电路
driver gear	从动齿轮
driving rod	传动杆
dropout fuse	跌落式熔断器
dross trap	扒渣
dry-type air-core reactor	干式空心电抗器
dye check/dye inspection	差色检查
earth potential	地电势（位）
stray current	杂散电流
edge pressure	边侧压力
eddy current	涡流
ejector pins/locking bolt/ram/throw-out	顶杆
electric charge	电荷
electric constant	电常数
electric field	电场
electric flux density	电通密度
electric shear	电剪
electrical degree	电度
electrical field configuration across the contacts and to earth	断口两端和对地的电场结构
electrochemical fluorination	电化氟化法
electrode	电极
electro-deposition	电沉积

electrodynamic force	电动力
electromagnet	电磁铁
electromagnet coil for closing	合闸电磁铁线圈
electromagnetic wave	电磁波
electromagnetic [solenoid] valve/electrovalve	电磁阀
electromagnetically induced current	电磁感应电流
electromagnetically radiation	电磁辐射
electron beam welding machine	电子束焊机
electron capture detector/electron-capturing detector	电子捕获检漏器
electro-slag welding machine	电渣焊机
electrostatically induced current	静电感应电流
encapsulating compound	包封胶
energizing circuit	激励电路
envelope curve	包络线
equivalent series resistance	等效串联电阻
erection/mounting/installation	安装
explosive gases	爆炸性气体
exponentially decaying d.c. recovery voltage	按指数衰减的直流恢复电压
Faraday rotation angle	法拉第旋转角
fast-front overvoltage	快波前过电压
ferrofluid	磁流体
fibres of dust setting on the insulator	存积在绝缘子上的粉尘纤维
file	锉刀
final installation inspection	安装竣工检验
flashover voltage	闪络电压
floating potential	漂移电位
forming machine	成型机
fracture	断裂(口)
frequency converter	变频器
frequency doubler	倍频器
full three-phase short-circuit current	满负荷的三相短路电流

续表

full voltage stress	全电压负荷
fuse/fuse wire	保险丝
fused short-circuit current	熔断器限制的短路电流
fusible (permanent backing)	可熔（保留垫板）
galvanised steel support structure	镀锌钢支架结构
galvanizing	镀锌
gas evolving circuit-breaker	产气断路器
gas insulated combined switchgear	复合封闭开关设备
gas-filled switchgear	充气式金属封闭开关设备
gasket groove	垫圈槽
gear housing/gear-box	齿轮箱
general assembly of earthing blade's single pole	地刀单极总装配
generation of electric energy	发电
glow discharge	辉光放电
grading capacitor	分压电容器
half a pole	半极
half control	半控
half-life	半衰期
half-wave/half-cycle/loop	半波
hammer crusher	锤式破碎机
hazard reducing device	防触电装置
heat conductivity	传热性
heat transfer coefficient	传热系数
heat transfer fluid	传热流体
hetero charge	混杂电荷
high voltage paper electrophoresis	高压纸电泳
high/low voltage prefabricated substation	高/低压预装式变电站
high-current interval[phase]	大电流阶段
high-frequency transient oscillation	高频瞬变振荡
high-pressure interlocking device	高压联锁装置
high-pressure ionization chamber	高压电离室

续 表

high-speed arc short-circuiting device	快速电弧短接装置
high-voltage alternating current switch-fuse combinations	高压交流负荷开关-熔断器组合电器
high-voltage interval	高电压阶段
holding temperature	保温温度
homo charge	纯号电荷
hybrid current-limiting circuit breaker	混合式限流断路器
imaging system	成像系统
impact bending test	冲击弯曲试验
impulse breakdown	冲击击穿
impulse equivalence	脉冲当量
impulse voltage generator	冲击电压发生器
impulse voltage withstand test	冲击电压耐受试验
inclusion body	包含体
increased operating frequency circuit-breaker	频繁操作断路器
indirect over-current release	间接过电流脱扣器
inductive breaking current	电感开断电流
inductively coupled plasma	感应耦合等粒子体
initial conversion ratio	初始换算比
initial transient recovery voltage envelope	起始瞬态恢复电压包络线
insertion current	插入电流
inspecting determining the leakage nature/qualitive leakage inspection	定性检漏
instant of contact separation	触头分开瞬间
instantaneous peak current	瞬时峰值电流
institude of electrical and electronics engineers	电气及电子工程师学会(美国)
insulated feeder	保温冒口
insulation between sections/section insulation	段间绝缘
interaction period around current zero	电流零点前后的相互作用期间
inter-electrode voltage	内电极电压
intermediate circuit	中间电路
interpass temperature	层间温度

续 表

interpolation value	插入值
interrupters in series	串联断口
interrupting capability/breaking capacity	开断能力
ionized gas	电离气体
isolator	刀闸（隔离开关）
keep relay	保持继电器
lagging/thermal insulation	保温层
laminer flow control	层流控制
lap welding	搭接焊
large cable network	大型电缆网络
large scale integrated circuit	大规模集成电路
large size boiler	大型锅炉
lathe	车床
layer insulation/insulation between layers	层间绝缘
lead coated metal	包铅金属
leakage current	泄漏电流
let-go current	摆脱电流
lethal voltage	致命电压
lightning impulse withstand voltage	雷电冲击耐受电压
line-charging breaking capacity	线路充电开断能力
liquid-filled combinations	充液式组合电器
liquid-filled switchgear	充液式开关设备
load-breaking switch/load-switch/switch	负荷开关
local selector	近控选择开关
lock valve	保险阀
locked-rotor torque	堵转转矩
locking bolt	闭锁螺栓
lock-out system	闭锁系统
long pre-arching time	长预弧时间
long unloaded line	长空载线路
loop circuit	闭回路

loop control system	闭环控制系统
loop duration	半波持续时间
low ohmic parallel resistor	低值并联电阻
low pressure(lp) circuit	低压回路
low-voltage apparatus	低压电器
lumped circuit	集中参数电路
magnetic blow-out circuit-breaker	磁吹断路器
magnetic field sensor	磁场(感应)传感器
magnetic overload relay	磁过载继电器
magnetizing inrush current	冲击励磁电流
magneto-starter (for ac motor)	磁力起动器(用于交流电机)
magnitude of fault current	故障电流量
make-proof earthing switch/high-speed earthing switch	快速接地开关
matching circuit	匹配电路
measuring current transformer	测量用电流互感器
measuring impedance	测量阻抗
mechanical back-to-back test	对拖试验
medium voltage electric power system	中压电力系统
minimum breakdown voltage	最小击穿电压
monostable relay	单稳态继电器
motor actuated pump	电动泵
motor pump group	电机油泵组
motor-circuit	电动机回路
mounting structure or spacing	安装结构或间距
multi-threshold amplitudes	多阈值幅值
negative phase-sequence impedance	负序阻抗
neutral voltage shift	中性点电压漂移
nominal value	标称值
nominal voltage	标称电压
non-draining cable	不滴流电缆
normal pressure	常压

续 表

normal/standard current rating/rated current	标准电流额定值
normal/standard short duration power-frequency withstand voltage	标准短时工频耐受电压
normal［standard］switching impulse	标准操作冲击
normally closed contact	常闭接点
nozzle control unit	喷口控制装置
nuclear magnetic double resonance	核磁双共振
number of interrupters in series	串联灭弧室数
oil-filled reactor	充油的电抗器
one-half cycle	半个周波
open type circuit-breaker	敞开式断路器
open type composite apparatus	敞开式组合电器
open-type switchgear	敞开式开关设备
operating torque	操作力矩
operating voltage/operational voltage/service voltage	工作电压
operation counter's circuit	动作计数器回路
operation situation of power network	电网运行情况
operational grade	操作组段
optical fiber cable	光纤电缆
order for goods/order form/order	订货单
oscillating current	振荡电流
overcurrent	过载电流
overload time relay	过载限时继电器
parallel (switching/shunt) resistor	并联电阻
parallel breaking resistor	并联开断电阻
parallel capacitor bank	并联电容器组
parallel circuit	并联电路
parallel circuits injection circuit	并联电流引入回路
parallel connected grading capacitor	并联均压电容器
parallel-to-serial converter	并联-串联变换器
partial discharge extinction voltage	局部放电熄灭电压

续　表

percent conductivity	百分数导电率
percentage impedance	百分阻抗
permeability	磁导率
phase converter	变相器
phase to phase capacitance of connection	相间联接电容
phase-sensitive voltmeter	相敏电压计
piezoelectric transition	压电跃变
piping and instrument diagram	配管及仪表图
plasma flow/plasma jet	等离子流
plate gauge (shaping plate)	板规(靠模)
platinum electrode	白金电极
plug	插头
plug jumper	插入跨接线
plug-in coil	插入式线圈
plug-in termination	插塞式终端
plug-in type termination	插入式电缆终端
point of current infeed	电流引入点
point-out-wave of contact separation	触头分离相位
pole closed	闭合相
poles in series	串联相
polygonal voltage	边电压
polyphase system	多相系统
portable electric tool	便携式电动工具
post-arc conductivity	弧后导电性
potential drop	电位降
power sources structure	电源结构
power-frequency voltage tests on the main circuit	主回路工频电压试验
power-operated mechanism/power operating device	动力操动机构
pressure in gauge/gauge pressure	表压
prevent disruptive discharge to earth	防止对地击穿放电
principle lay-out of the circuit	电路原理图

续表

prospective symmetrical current	预期对称电流
protection level of a protective device	保护装置的保护因数
protection power gap	保护电力间隙
protective circuit	保护电路
protective reactor	保护电抗器
protective relay	保护继电器
protective spark gap	保护火花间隙
protective transformer	保护用互感器
pump	泵
pump discharge control valve	泵出口控制阀
pump drive assembly	泵传动装置
pumping device	泵路装置
radio-frequency cable	射频电缆
rated supply voltage	额定电源电压
co-ordination table of rated values	额定值配合表
rated switching impulse withstand voltage	额定操作冲击耐受电压
ratio error	比差
ratio of current amplitude	电流幅值比
reactive power	无功
receptacle/socket/socket-outlet	插座[孔]
recharging device	储能装置
recharging time of the operating mechanism	操动机构的储能时间
recovery voltage	恢复电压
recovery voltage across circuit breaker	断路器的恢复电压
reference offset/zero offset/zero shift	零点偏移
relay for opening lockout/trip lock out relay	分闸闭锁继电器
relay for opening lockout at low oil pressure	低油压分闸闭锁继电器
relief/safety valve	安全阀
resetting current	复位电流
residual stress	残余应力
residual voltage	残余电压

续 表

restraint current	制动电流
retention index	保留指数
reverse current relay	逆流继电器
reverse current release	反向电流脱扣器
discard/refuse/scrap	报废
rod insulator	棒形绝缘子
routine test program	出厂试验大纲
rules for electric traction equipment	电力牵引设备规程
rupture	爆裂
safe and reliable operation	安全可靠运行
safe certification	安全认证
safe technical regulation	安全技术规程
safeguarding specifications	安全防护技术要求
safety extra-low voltage	安全特低压
safety margin	安全裕度
safety operation specifications	安全操作程序
safety precautions	安全注意事项
saturation	饱和
saturation factor	饱和因数
saturation magnetization	饱和磁化强度
scalar quantity	标量
scrapped product	报废品
screened cable	屏蔽电缆
secondary current	次级电流
secondary side of transformer	变压器副边
secondary winding	次级线圈
self arc diffusion of electrode(SADE)	电极的电弧自扩散
semiannual inspection and test	半年度检验
semi-automatic	半自动的
semi-automatic welding	半自动焊
semiconductor	半导体

续 表

semiconductor layer	半导体层
semi-dry method	半干法
semi-finished product	半成品
series capacitors	串联电容器
series connected accessories	串联附件
series impedance of the combinations	组合电器的串联阻抗
series inductance	串联电感
series voltage injection	串联电压引入法
series-parallel starting	串并联起动
servomotor	伺服电动机
sf6-filled switchgear	充 SF6 的开关设备
sheet backing	背板
sheets	板材
shield of moving contact	动触头屏蔽罩
shock current	触电电流
shop welding	车间焊接
short circuit interrupting capability/short-circuit breaking capacity	短路开断能力
short circuit of layers	层间短路
short connections to voltage transformer	电压互感器的短连接线
Short-Circuit Testing Liaison（STL）	短路试验协会
short-line fault unit test	近区故障单元试验
shunt capacitor	并联电容器
shunt closing/opening release	并联合/分闸脱扣装置
shunt reactor	并联电抗器
shunt reactor switching	并联电抗器合、分操作
shunt reactor winding	并联电抗器绕组
shunted condenser[capacitor]	并联电容器
silver-plated copper	镀银铜
single phase gas circuit breaker	单相气体断路器
single pole circuit breaker	单刀(极)断路器

续 表

single-break contact assembly	单断点触头组
single-conductor cable	单芯电缆
single-phase-to-earth fault current	单相接地故障电流
site supply voltage	现场电源电压
slab	板坯
slow-front overvoltage	缓波前过电压
small oil volume circuit-breaker with live tank	带电油箱少油断路器
solenoid operated air valve	电磁气动阀门
solenoid operated valve	电磁控制阀
solid conducting link	导电棒
space charge probing technique	空间电荷探测技术
spare auxiliary switch	备用的辅助开关
spare parts	备件
spark plug	火花塞
sparkover of spark-gap	火花间隙的放电
sparkover voltage of making gap	关合间隙放电电压
specific resistance/resistivity	电阻率
specific saturation magnetization	比饱和磁化强度
specified-time relay	定时限断电器
speed-varying motor	变速电动机
standard atmospheric conditions	标准大气条件
standard reference atmosphere	标准参考大气
standard switching (lighting) impulse withstand voltage	标准操作(雷电)冲击耐受电压
standard test finger	标准金属试指
standard voltage shape	标准电压波形
static terminal load force	端子静负荷力
stationary electric contact	固定电接触
step-down transformer	降压变压器
storing energy and closing	储能合闸
stroke limitation	冲程限位器
submarine cable	海底电缆

续 表

substation	变电站
substation running mode	变电站运行方式
subtransient reactance	次瞬态电抗
sub-transmission network	次输电网
suction fan	抽风机
supercritical pressure unit	超临界压力机组
superluminescent diode(SLD)	超高度发光二极管
surface acting agent	表面活性剂
surface charge density	表面电荷密度
surface polarization	表面极化
surge absorber	冲击脉冲吸收器
surge arrester	避雷器
swamping resistance	扩程电阻
switch for closing/circuit-closing switch	合闸开关
switch for opening circuit's direct current supply	分闸回路直流电源开关
switch-disconnector	隔离式负荷开关
switching impulse voltage test	操作冲击电压试验
switching impulse withstand voltage	操作冲击耐受电压
switching over-voltage	操作过电压
switching-on	接通(电源)
take-over current	交接电流
tap	分接头
tap position	档位
temperature-keeping heater	保温加热器
tensile of vertical test	垂直拉力试验
test piece [object] of high capacitance	大电容试品
the self-generated transient overvoltage due to switching	开关操作自身产生的瞬态过电压
the series parallel combinations	串并联组合件
the test circuit breaker	被试断路器
the vicinity of the live part	带电部件附近
thermal insulation material	保温材料

续 表

thermally stimulated current (TSC)	热受激电流
thin film hybrid circuit	薄膜混合电路
thin film mercury electrode	薄膜汞电极
through fault current	穿越性故障电流
tightness of gas-filled compartment	充气隔室的密封
to evacuate/evacuation	抽真空
transducer[sensor]	传感器
transfer arm	传送臂
transformer	变压器
transformer magnetizing current	变压器磁化电流
transformer ratio/ratio of transformation	变比
transformer secondary terminal	变压器二次端头
transient overvoltage phenomena	暂态过电压现象
transient recovery voltage(TRV)	瞬态恢复电压
tree(surface) discharge	表面树枝状放电
triggering electrode	触发电极
triggering of thyristor	硅可控整流器触发
triggering pulse	触发脉冲
turbidity of water at outlet	出水浊度
turning speed	车削速度
two-parameter reference lines	两参数参考线
two-terminal-pair network	二端对网络
ultra-high-voltage circuit breaker	超高压断路器
ultrasonic pressure loss	超声波压力降
under-voltage opening release	欠压分闸脱扣器
unintentional rupture of bursting disk	爆破片的误爆破
urban area network	城域网
vacuum switching device	真空开关电器
value of the three-phase symmetrical current	三相对称电流值
value of the trapped charge	残留电荷值
variable autotransformer	可变自耦变压器

续 表

variable flux voltage variation	变磁通调压
varistor element	变阻元件
vertical centreline	垂直中心线
very-fast-front overvoltage	陡波前过电压
vicinity of centres of generation	发电站中心附近
voltage	电压
voltage contribution	电压增量
voltage grading capacitor/condenser/field-distributing capacitor	均压电容器
voltage source voltage	电压源电压
voltage-to-pulse converter	电压-脉冲变换器
wagging vibration	摆动振动
watt loss/power loss	功率损耗
welding with weaving; weave bead welding	摆动焊
wheeling power loss	过网线损
withstand voltage across contacts	触头间耐受电压
withstand voltage to earth	对地耐受电压
woven resistor	编织电阻器
wrench/spanner	扳手
zero sequence capacitance	零序电容

3. 工程管理

acceptance	验收
activity duration estimating	活动历时估算
actuals	现货
ACWP = Actual Cost of Work Performed	已执行工作实际成本
AD = Activity Description	活动描述/说明
AF = Actual Finish Date	实际完成日期
AOA = Activity-on-Arrow	箭线网络图(双代号网络图)
AON = Activity-on-Node	节点式网络图(单代号网络图)
AS = Actual Start Date	实际开始日期

续　表

baseline	基线
change control	变更控制
Change Control Board (CCB)	变更控制委员会
close out	竣工
communications planning	沟通规划
concurrent engineering	并行工程
constraints	约束条件
contingencies	意外费用
contingency	应急预算
contingency allowance	意外准备金
contingency reserve	意外储备
contract administration	合同管理
contract close-out	合同收尾
control chart	控制图
cost benefit analysis	成本收益分析
cost budgeting	费用预算
cost estimating	费用概算
CPI = Cost Performance Index	费用绩效指数
CPM = Critical Path Method	关键路线法
CV = Cost Variance	费用偏差
DD = Data Date	数据日期
deliverable	交付物
dependency	依赖关系
DU = Duration	延续时间
dummy activity	虚拟活动
duration compression	延续时间压缩
EAC = Estimate At Completion	在完成时的费用估算
earned value analysis	挣值分析
EF = Early Finish Date	最早完工日期
effort	人工
ES = Early Start Date	最早开始日期

续表

ETC = Estimate To Complete	到完成时的估算
EV = Earned Value	挣值法
event-on-node	单节点事件图
exception report	例外报告
FF = Finish-to-Finish	完成到完成关系
FF = Free Float	自由时差
float	时差/机动时间/浮动时间
forward pass	顺推计算法
FS = Finish-to-Start	完成到开始关系
functional manager	职能经理
Gantt chart	施工进度表（甘特图）
graphical evaluation and review technique	图解评审技术
IFB = Invitation for Bid	邀标
integrated cost/schedule reporting	进度综合报告
lag	滞后量
line manager	产品经理
mitigation	风险缓解
PC = Percent Complete	完成百分比
performance reporting	执行报告
PERT chart	计划评审技术图
PERT = Program Evaluation and Review Technique	计划评审技术
PMBOK = Project Management Body of Knowledge	项目管理知识体系
precedence relationship	优先关系
predecessor activity	紧前工作
procurement planning	采购规划
project champion	项目领导
project charter	项目许可证
project human resource management	项目人力资源管理
project in controlled environment	受控环境中的项目
project life cycle	项目生命周期
project network diagram	项目网络图

续　表

project plan development	项目计划开发
project plan execution	项目计划实施
project procurement management	项目采购管理
project quality management	项目质量管理
project risk management	项目风险管理
project scope management	项目范围管理
QA = Quality Assurance	质量保证
QC = Quality Control	质量控制
RAM = Responsibility Assignment Matrix	责任分配矩阵
RDU = Remaining Duration	剩余持续时间
resource-limited schedule	资源约束进度计划
RFP = Request for Proposal	请求建议书
RFQ = Request for Quotation	请求报价单
risk identification	风险识别
risk response control	风险应对控制
schedule	工期
scope baseline	范围基准计划
scope verification	范围验证
slippage	延误,逾期
solicitation	询价
SOW = Statement of Work	工作说明
SPI = Schedule Performance Index	进度执行指数
staff acquisition	工作人员招募
stakeholders	利益相关者
SV = Schedule Variance	进度偏差
target schedule	目标时度计划
TC = Target Completion Date	目标完成日期
time-scaled network diagram	时标网络图
TQM = Total Quality Management	全面质量管理
WBS = Work Breakdown Structure	工作分解结构
workaround	权变措施

参考文献

[1] 成虎,陈群.工程项目管理[M].4版.北京:中国建筑工业出版社,2015.
[2] 陈维政,等.人力资源管理与开发高级教程[M].3版.北京:高等教育出版社,2019.
[3] 丁士昭.工程项目管理[M].北京:中国建筑工业出版社,2009.
[4] 方华文.20世纪中国翻译史[M].西安:西北大学出版社,2008.
[5] 方梦之,范武邱.科技翻译教程[M].上海:上海外语教育出版社,2015.
[6] 冯梅,刘荣强.英汉科技翻译[M].哈尔滨:哈尔滨工业大学出版社,2000.
[7] 高然.能源电力翻译理论与实践[M].天津:天津科学技术出版社,2018.
[8] 韩礼德,哈桑.英语中的衔接[M].伦敦:朗曼出版社,1976.
[9] 何其莘,仲伟合,许钧.科技翻译[M].北京:外语教学与研究出版社,2012.
[10] 胡钋,华小梅.电力工程英语综合教程[M].北京:中国电力出版社,2011.
[11] 刘健,等.电力英语阅读与翻译[M].北京:中国水利水电出版社,2012.
[12] 刘健.电力英语阅读与翻译[M].北京:中国水利水电出版社,2022.
[13] 刘然,包兰宇,景志华.电力专业英语[M].北京:中国电力出版社,2004.
[14] 马祖毅.中国翻译史(上卷)[M].武汉:湖北教育出版社,1999.
[15] 马工程管理学编写组.管理学[M].北京:高等教育出版社,2019.1.
[16] 任淑平,何晓月.工程英语翻译理论与实践研究[M].沈阳:东北大学出版社,2022.
[17] 史澎海.工程英语翻译[M].西安:陕西师范大学出版社,2011.
[18] 孙昌坤.实用科技英语翻译[M].北京:对外经济贸易大学出版社,2013.
[19] 孙建光,李梓.工程技术英语翻译教程[M].南京:南京大学出版社,2021.
[20] 孙有中,任文,李长栓.高级汉英笔译教程[M].北京:外语教学与研究出版社,2022.
[21] 王宏志.翻译史研究(2013)[M].上海:复旦大学出版社,2013.
[22] 武力,赵拴科.科技英汉与汉英翻译教程[M].西安:西北工业大学出版社,2000.
[23] 谢龙水.工程技术英语翻译导论[M].北京:北京希望电子出版社,2015.
[24] 谢天振,何绍斌.简明中西翻译史[M].北京:外语教学与研究出版社,2013.
[25] 严俊仁.新英汉科技翻译[M].北京:国防工业出版社,2010.
[26] 闫文培.实用科技英语翻译要义[M].北京:科学出版社,2008.
[27] 杨士焯.西方翻译理论[M].厦门:厦门大学出版社,2018.
[28] 杨太华,等,电力工程项目管理[M].北京:清华大学出版社,2017.3.
[29] 张干周,郭社森.科技英语翻译[M].杭州:浙江大学出版社,2015.

［30］张素贞,刘晓艳.新能源专业英语:New energy in English[M].北京:化学工业出版社,2014.
［31］张艺,容庆.创业管理[M].北京:清华大学出版社,2020.
［32］赵萱,郑仰成.科技英语翻译[M].北京:外语教学与研究出版社,2006.
［33］赵玉闪,吕亮球,等.电力英语互译理论与实践研究[M].保定:河北大学出版社,2012.
［34］左广明,李纯.科技文体翻译教程[M].武汉:武汉大学出版社,2012.